Pre-Geometry Brain Teasers

Author:

Sylvia J. Connolly

Introduction

Geometry is the study of space. Space exists in the third dimension. This is the dimension in which we live. There is length, width, and depth to our reality. Two-dimensional space exists only in the mind—length and width (no depth). One-dimensional space also exists in the mind—a line that has direction but no thickness. No dimension is almost impossible to imagine, but it is a point having no measurements. All dimensions other than the third have to be imagined. Students who have vivid imaginations are wonderful candidates to learn geometry. If you have students who like to draw and are visually inclined, geometry should be fun for them.

The whole organization of geometry is similar to the rules of a game. You cannot play the game unless you know the rules. Learning the rules makes you a better game player. Geometry is the same way.

Pre-Geometry Brain Teasers will take the main rules of geometry and create problems that will reinforce the rules. This book is not meant to teach geometry but to supplement the textbook with activities that encourage students to expand their thinking. The exercises are also meant to be a springboard from which you as the teacher can think of other exercises that engage the class. Have the students use their previous work. Remind them that when one works in the business world, he or she often stores previous work for future reference. Teaching students to do the same is great training for the future.

Contributing Author:

Karen J. Goldfluss, M.S. Ed.

Editor:

Stephanie Buehler, M.P.W., M.A.

Teacher Created Materials, Inc.

6421 Industry Way

Westminster, CA 92683

Made in U.S.A.

ISBN-1-57690-218-8

Graphic Illustrator:

James Edward Grace

Cover Artist:

Larry Bauer

Chris Macabitas

Table of Contents

Tables of Contents (cont.)

How to Use This Book

The brain teaser exercises in this book require little preparation for the student or teacher. Each exercise begins with a set of student directions. Where necessary, teacher directions and/or solutions are provided on pages marked "Teacher Page" at the beginning of each section. As your students work through exercises in this book, keep the following information in mind.

1. Read the directions. Take some time and think about how this relates to all of geometry. If students do the exercises differently from the way in which you or I would do them, don't mark them incorrect. Try to figure out how they thought this out. Try to understand how they think.
2. When making lines, use a sharp pencil and a straight edge. Don't use the side of a book or a card.
3. Geometry is related to art. Some of your students will create beautiful patterns that you will want to display. Make certain that everyone is encouraged to participate in a classroom display.

Note: Where possible, metric conversions have been provided. If metric measurement is not given, add or substitute the metric conversion information before reproducing a page.

The tools needed include the following:

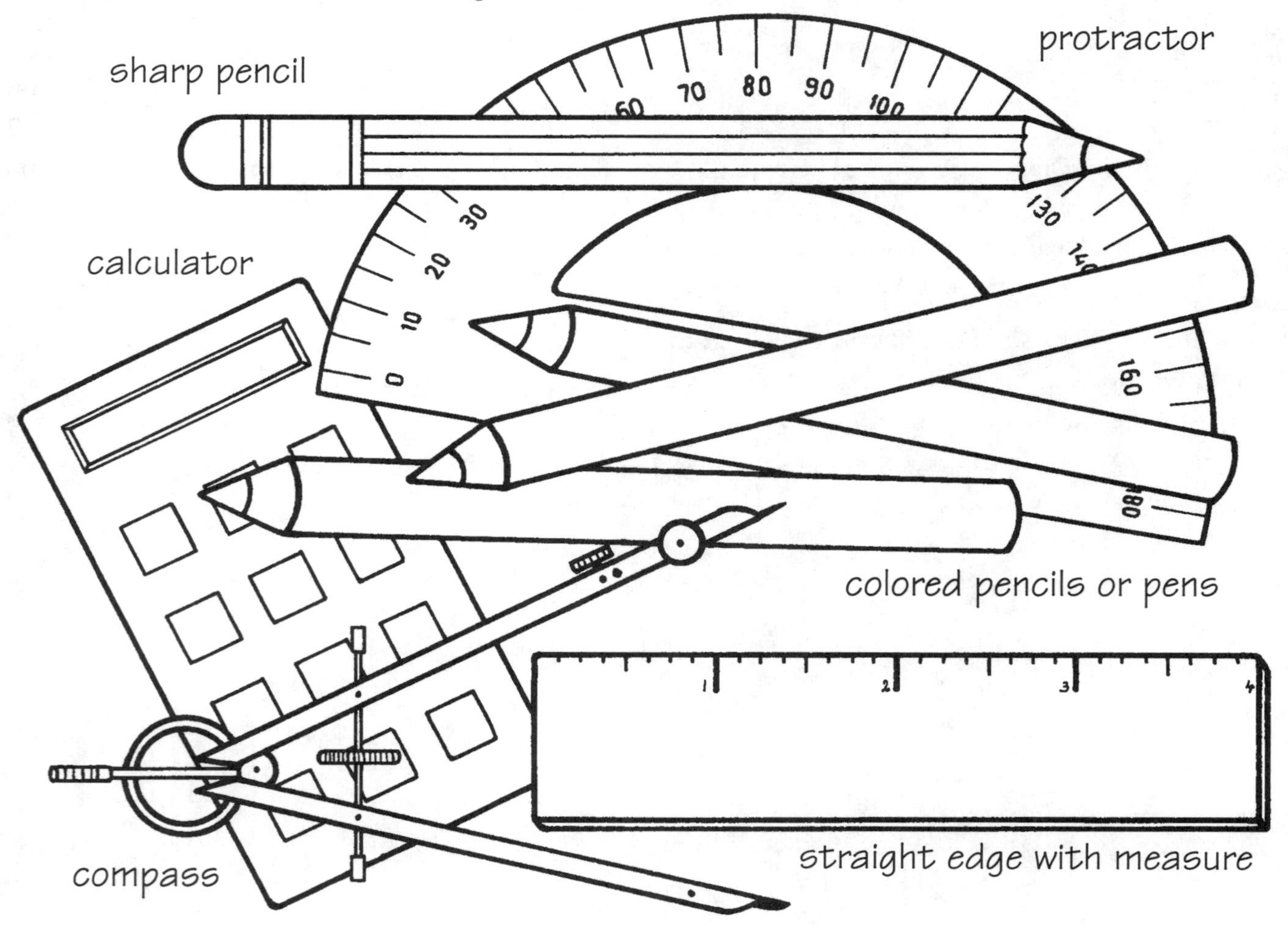

Terms Used in Geometry

The following terms are provided for easy reference as you complete the pages of this book.

acute triangle—a triangle in which all the angles are less than 90°
adjacent angles—angles having one common side between them and one common vertex
angle—a figure consisting of two lines that meet at a common point
area—a measurement derived from the number of square units in a figure
axiom—a mathematical principle that can be demonstrated but not proved
base—a figure's top or bottom
bisector—a line dividing a geometric figure into two equal parts
chord—a line segment in which the endpoints fall on the circle
circle—a curved line formed by a set of points that are equidistant from a center point
circumference—the distance measure of a circle
congruent figures—figures that fit on one another
cube—a rectangular solid with six equal faces
diagonal—a line segment that joins two nonconsecutive vertices in a polygon
diameter—a straight line radiating from a point on a circle, through its center, and to another point on the circle
equiangular triangle—a triangle having three equal angles of 60°
equilateral triangle—a triangle having three equal sides
geometry—the area of mathematics that deals with spatial relationships
horizontal line—a straight line that is parallel to the plane of the horizon
hypotenuse—in a right triangle, the side opposite the right angle
isosceles triangle—a triangle having two equal sides
line—an infinite number of points in a straight arrangement (A line has length, but no thickness or width.)
obtuse triangle—a triangle in which an angle is greater than 90°
pi—a ratio derived by dividing the circumference of a circle by its diameter ($\pi = c/d$)
point—an object that shows the location or position of the intersection of two lines
polygon—a geometric figure of three or more joined line segments and three or more angles
postulate—a geometric principle that can be demonstrated but not proved
radius—a line joining the center of a circle to any point on the circle
rectangle—a four-sided plane figure with four right angles
right angle—an angle measuring 90°
scalene triangle—a triangle in which there are no equal sides
solid geometry—the area of geometry that deals with three-dimensional objects
square—a type of parallelogram having four equal sides and four right angles
theorem—in geometry, a mathematical idea that can be proved sometimes using axioms or postulates
vertex—the point where two lines meet (When referring to more than one vertex, the term vertices is used.)
volume—the amount of space, measured in cubic units, in a solid

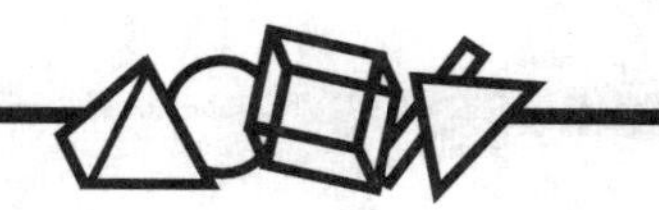

Formulas Used in Geometry

The following formulas are provided for easy reference as you complete the pages of this book.

Area		
Figure	**Formula**	**Using the Formula**
Square	$A = s^2$	Square one side or multiply the length of one side by another.
Rectangle	$A = lw$	Multiply a length by a width.
Parallelogram	$A = \frac{bh}{2}$	Find the product of the length of the base and the height.
Triangle	$A = \frac{bh}{2}$	After multiplying the length of the base by the height, divide the product by 2.
Trapezoid	$A = \frac{(b_1 + b_2)h}{2}$	First, add the lengths of the parallel bases. Then, multiply the sum by the height and divide this product by 2.
Circle	$A = \pi r^2$	Find the square of the radius and multiply that number by 3.14 or 22/7 ($\pi \approx 3.14$ or 22/7). Note: If given a diameter, divide the diameter by 2 to find the radius.

Volume		
Cube	$V = s^3$	Measure a side (s) and find the product of s to the power of 3.
Rectangular	$V = lwh$	Find the product of the length, width, and height.
Cylinder	$V = \pi r^2 h$	First find the area of the base of the cylinder, which is a circle. In the formula, this area is represented by πr^2. Then, multiply this area by the height of the cylinder.

Tools

If commercial measuring tools are not available, you may wish to use these. Copy the patterns on heavy paper and cut them out. A review of how to use a protractor is provided on page 8.

Protractor

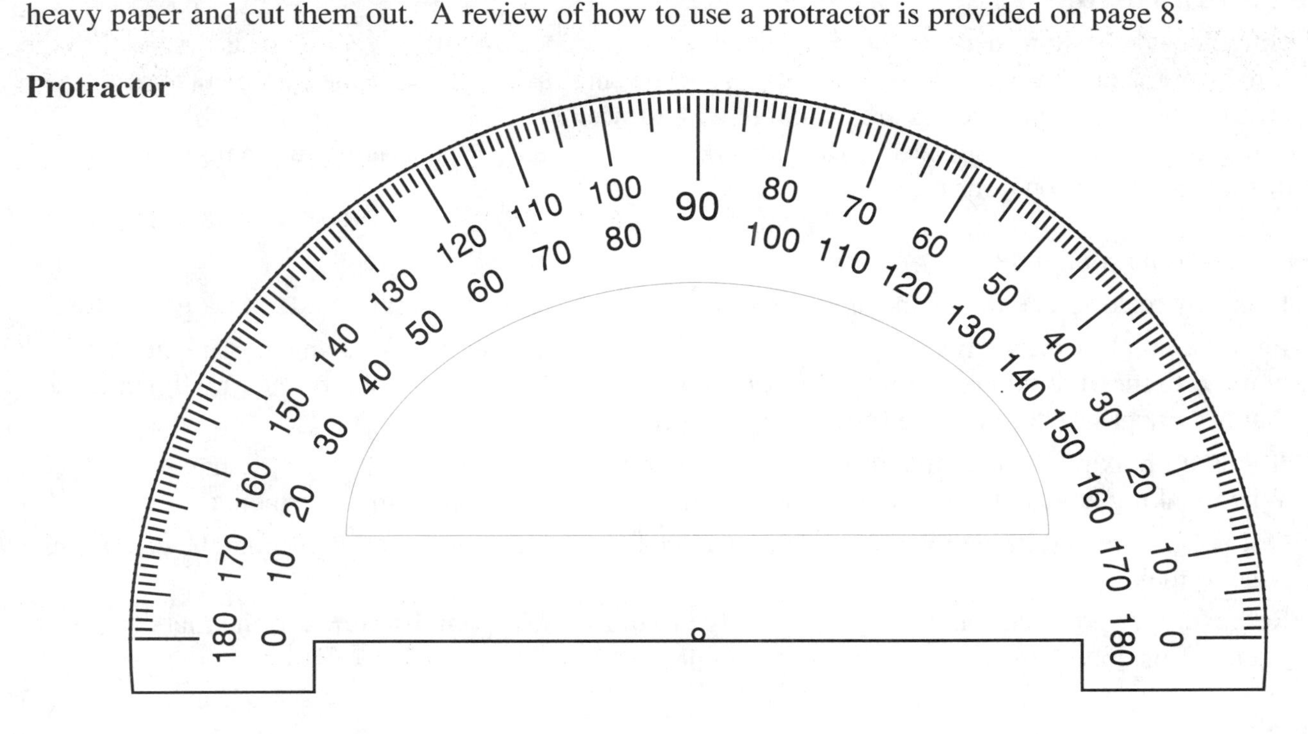

Inch Ruler

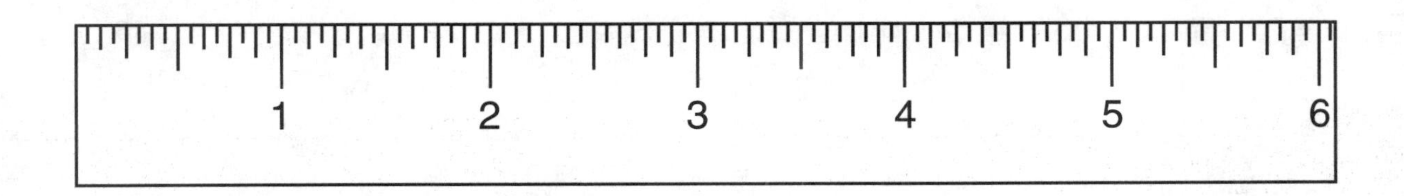

Centimeter Ruler

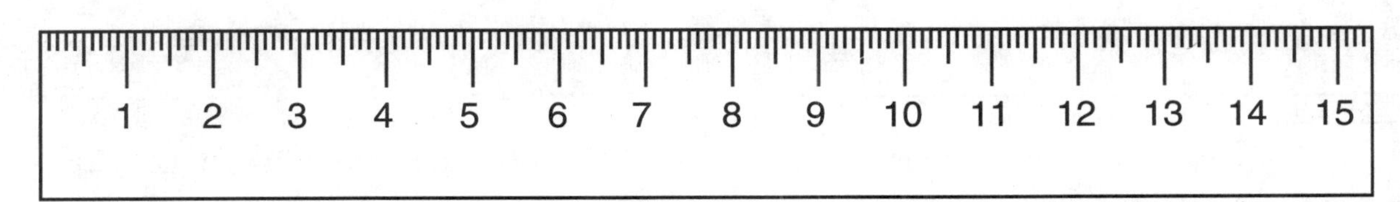

Tools (cont.)

Using a Protractor

The protractor is a basic measuring tool in geometry. The proper use of the protractor is essential to the correct measurement of angles. Follow these steps to be sure that your angle measurements are accurate. (Note: Many protractors are available with movable straightedge attachments that make measuring and drawing an angle easy. The steps below are based on the assumption that you do not have this added feature on your protractor.)

Steps for Making an Angle

1. Begin by making one ray of the angle. (You can use a ruler or the straight edge of a protractor).
2. Find the point or mark on your protractor that represents the vertex of the angle. Line up this point with the endpoint of the ray you just drew. Be sure the base of the protractor is lined up so that the ray passes through the 0° mark on the protractor.
3. If you are making an angle that opens on the right side, use the lower set of angle measurements. When making angles that open on the left, use the upper set of angle measurements.
4. Once you have determined which scale to use, make a mark (point) that represents the angle you wish to make.
5. Remove your protractor and use a straightedge to draw a ray from the vertex to the mark you just made. This completes the two rays of the angle.

Examples

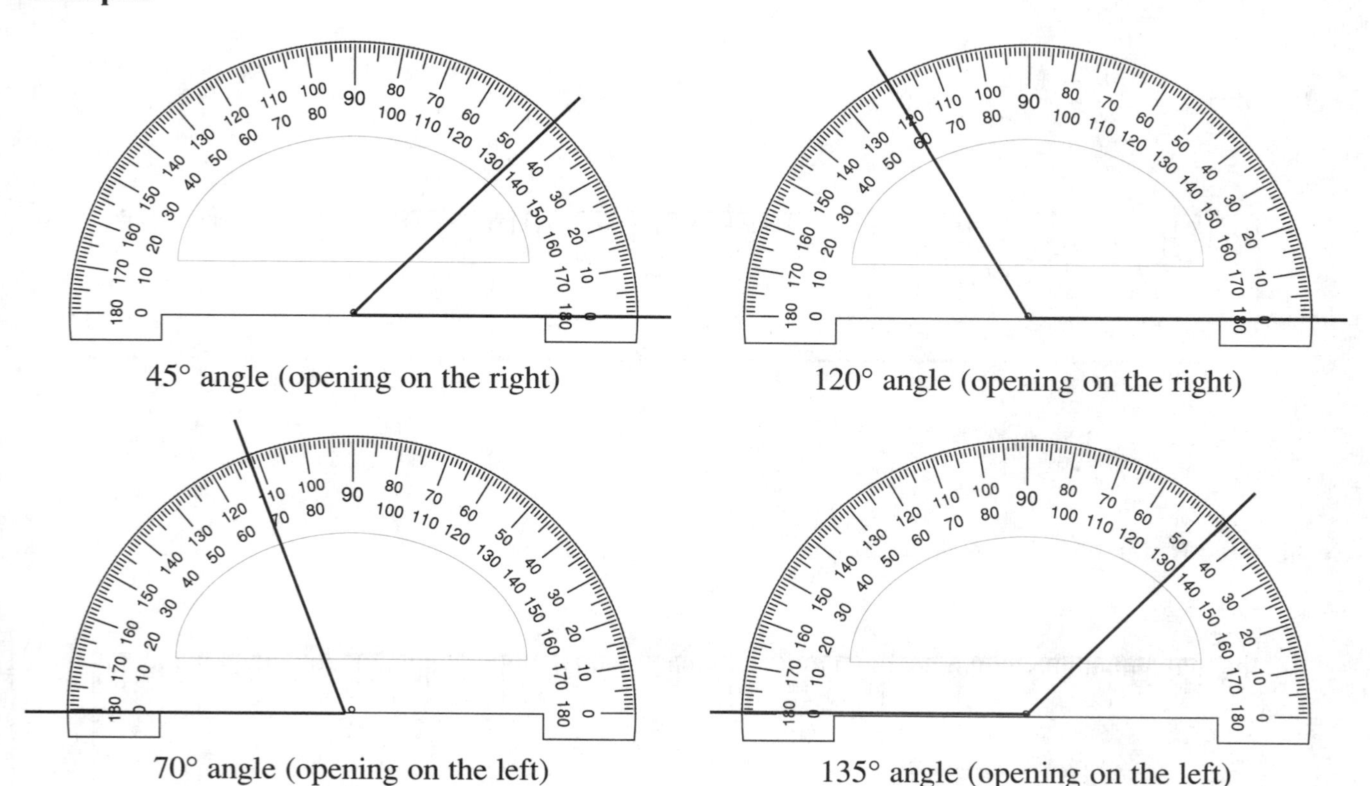

45° angle (opening on the right)

120° angle (opening on the right)

70° angle (opening on the left)

135° angle (opening on the left)

Teacher Pages: Points, Lines, and Patterns in Geometry

The following information corresponds to the exercise pages in this section. Where applicable, teacher information that may be helpful to the students' understanding of the exercises is provided. For easy reference, the page numbers and titles of the exercises are given.

The exercises in this section are designed to get students thinking about points, lines, space, and patterns. Introduce students to the following geometric terms as you work through this unit. These words will be used in this section and throughout the book.

- **Point**—This is the basic unit in geometry. The definition of a point as it is used in geometry is an object that shows location and is without dimension. Although a point is smaller than the smallest dot one can make, students should realize that they will be using a dot made by a pencil to represent a point. The dot is only a physical representation, or model of a point. This idea may seem very abstract to students, but it is this use of abstract thought that allowed the definitions and rules of geometry to be developed. Because geometry can be very abstract to students, it is also important to allow students time to reflect on what they have learned.
- **Line**—A line is an infinite number of points in a straight arrangement. A line has length but no thickness or depth. It extends forever in two directions.
- **Plane**—A plane is a flat surface that extends out indefinitely in all directions. Have students imagine the surface of the chalkboard or desktop as part of a plane that stretches out in all directions indefinitely.

Page 12: Connecting Points

This exercise is good for students to see that their work can sometimes be very different from their neighbor's and still be correct. Have all students display this paper on the board. Then discuss as a class the differences and similarities. Some students like to color in their names. The last letter and the first should be connected in order to make a complete design. This will also introduce the visual picture of a polygon.

Page 13: Line Segments Using Friends' Names

Make certain that the student is connecting the points with a straightedge because it is a temptation to do this exercise freehand. Reinforce that this is what a draftsperson or an architect does when making lines.

Pages 14 and 15: Copying a Line Segment

Line segment measurements on page 15 are as follows: $\overline{AB} = 4''$, $\overline{CD} = 2''$, $\overline{EF} = \frac{1}{2}''$, $\overline{GH} = 3''$, $\overline{IJ} = 2\frac{1}{2}''$, $\overline{KL} = 4\frac{3}{4}''$, $\overline{MN} = \frac{1}{2}''$, $\overline{OP} = 3\frac{1}{4}''$, $\overline{QR} = 1\frac{1}{4}''$, and $\overline{ST} = 2\frac{1}{4}''$. Any arrangement of line segments can be used as long as the result is 4 sides, each 6" (15 cm) long.

Page 16: More Line Segments

Note that the line segments are not in one direction. The points become endpoints of two line segments. The language used is that they share the same endpoint. Another way to prove that the line segments add up to line segment EZ is to measure each carefully and add them. Then measure line segment EZ. The transitive property says that if $a = b$ and $b = c$, then $a = c$. Euclid said, "The whole equals the sum of its parts."

Teacher Pages: Points, Lines, and Patterns in Geometry (cont.)

Page 17: Simple Map Copying

Getting students to work a compass is a way not only to use construction in geometry, but also a way to measure. Some students will find this easy, while others will have problems. Have the students work in groups for this as they will help each other.

Pages 18–20: Perspective

Make sure students understand the terms introduced on these pages. If necessary, review parallel lines, diagonals, vertical and horizontal lines, vanishing points, and horizon lines. When students have completed the activities on these pages, they will have a better understanding of the role of geometry in everyday life.

Page 21: Pascal's Pattern

This page introduces students to Pascal's triangular array of numbers. There are several discoveries students can make about the pattern of numbers in this array. Students are challenged to complete several more rows of the triangle and to state observations about the interesting patterns found in the array. Learning to closely observe patterns and similarities is an important skill for students of geometry.

Among the possible discoveries students might make about patterns are the following:

- The triangle is symmetrical, with the line of symmetry being the vertical line of numbers that appear in the center of the array.
- The numbers read the same from left to right as they do from right to left.
- The first diagonals in either direction are always 1.
- The second diagonal contains the counting numbers 1, 2, 3, 4, and so on.
- The sum of any two adjacent numbers can be found in the row directly below and in between the two numbers.
- Students may come up with several more patterns. Any supported response should be accepted.
- Challenge students to design other arrays similar to Pascal's Triangle.

Teacher Pages: Points, Lines, and Patterns in Geometry (cont.)

In this section, students will take a closer look at points and extend their thinking about lines and line segments. Introduce or review the following information with students before they begin the brain teaser pages.

Definition of a line: a set of points having direction

This means that the line has one dimension. The number of points can be infinite and the points can go off into infinity. (Have students play with this idea.) It is easier to just put down two points and connect them with a straight line.

Share the following information with the students.

$\overline{AB}$: line segment AB

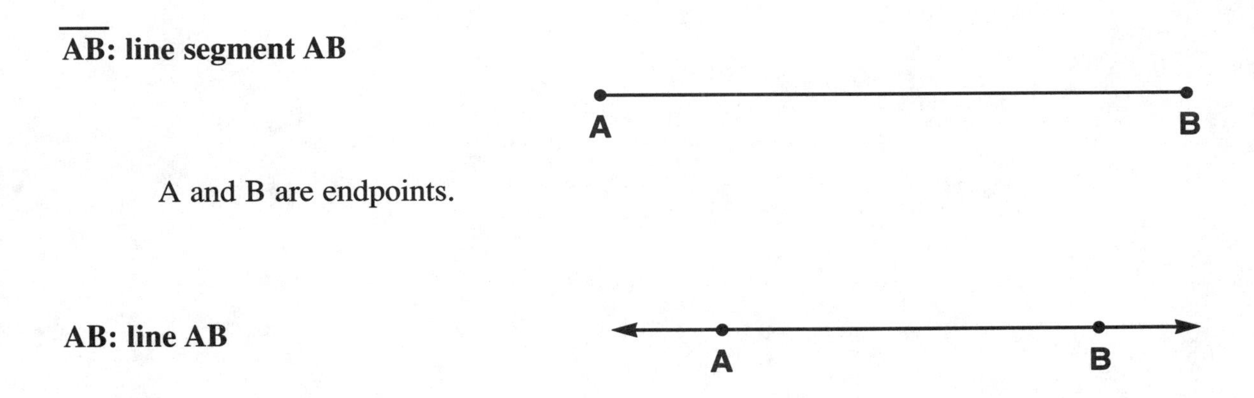

A and B are endpoints.

AB: line AB

There are no endpoints here, just a line through the points A and B.

$\overrightarrow{AB}$: ray AB

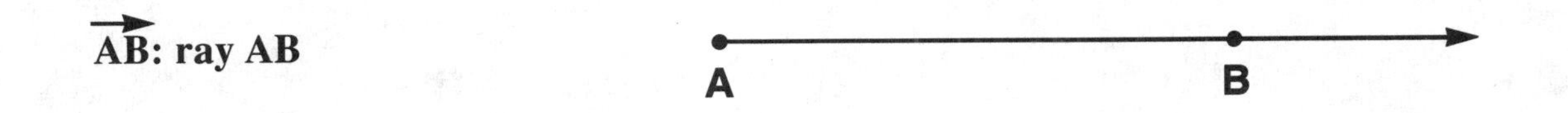

There is one endpoint, A. The line then goes through B to infinity.

$\overrightarrow{BA}$: ray BA

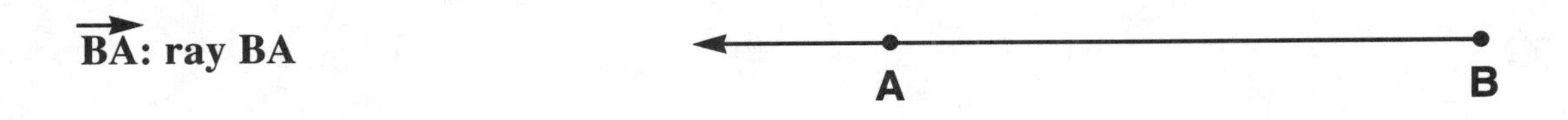

There is one endpoint, B. The line then goes through A to infinity.

Connecting Points

A point is a dimensionless object that shows location. The drawing of a dot represents a point so we have a way to communicate about it in reality. A point is usually labeled by a capital letter as shown below.

•A	point A, or A
This is read as "Point A" or simply "A".	

Directions: Connect the following points, using the letters in your name as endpoints. No one's drawing should look like any one else's line segment drawing. Use a straight edge to accurately connect the points and different colors of ink for your first, middle, and last names.

• A • B • C • D

• E • F • G • H • I

• J K • L • M • • N

O • P • Q • R • S •

T • U • V • W • X •

Y • Z •

Line Segments Using Friends' Names

Directions: Connect the points for the letters in your name in one color and the names of three additional people in three other colors. (If you have more than one of the same letter, count all of them as only one letter.)Connect the last letter with the first letter of each name to create a geometric shape. Use a ruler to accurately connect the points.

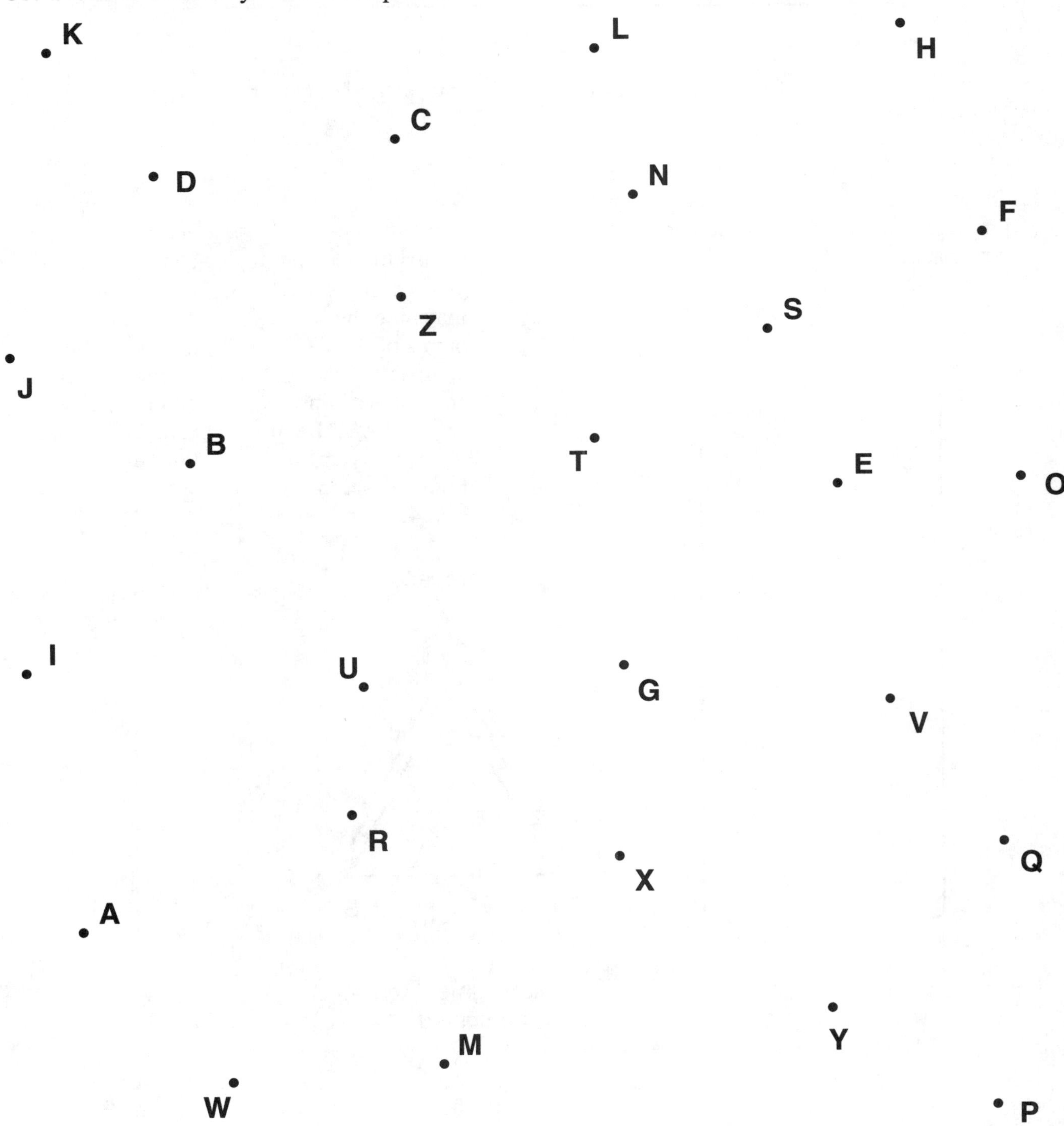

Copying a Line Segment

Directions: Use the steps below to learn how to copy a line segment. Practice copying line segment AB. When you think you are ready, complete page 15.

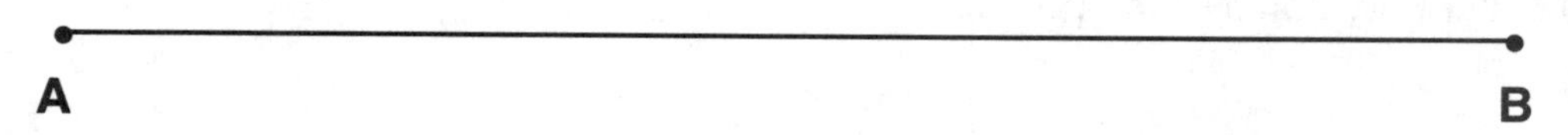

Copy this line segment. Follow steps 1–4 below.

1. Draw a vertical line segment on your paper.
2. Place a point on the line. Label it.
3. Measure the line segment to be copied by opening your compass and matching the points of the compass to the endpoints of line segment AB. Keep the compass open to this position for step 4.
4. Place the pointed end of the compass on your line at point A. Make a mark across the line segment to show point B. This is called copying a line segment.

A

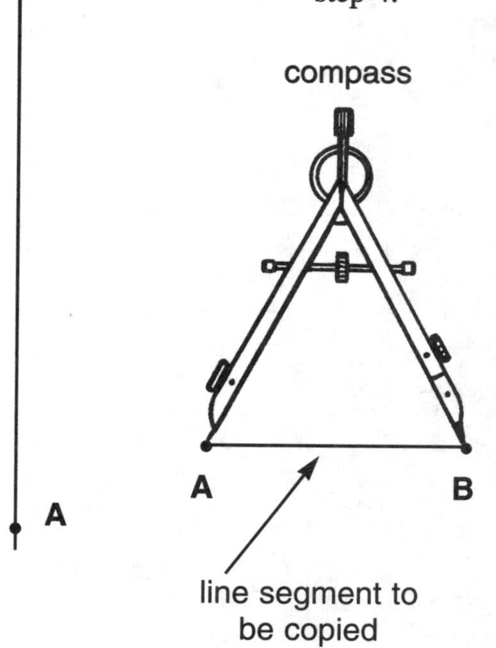

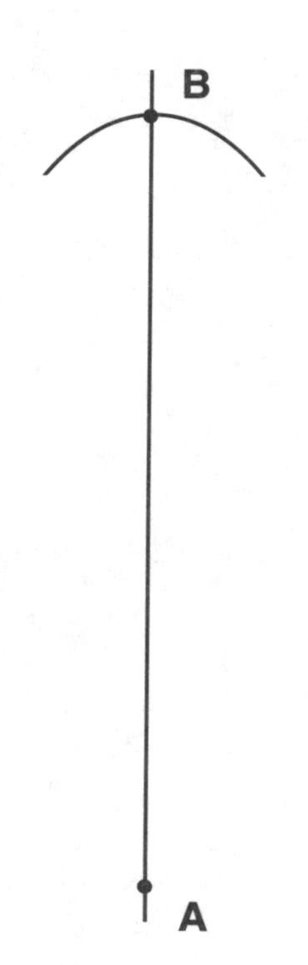

Copying a Line Segment (cont.)

Directions: Complete this exercise after you have reviewed the information on page 14. Measure the following line segments. Copy the line segments onto another piece of paper and connect them so that they make a 6" (15 cm) square.

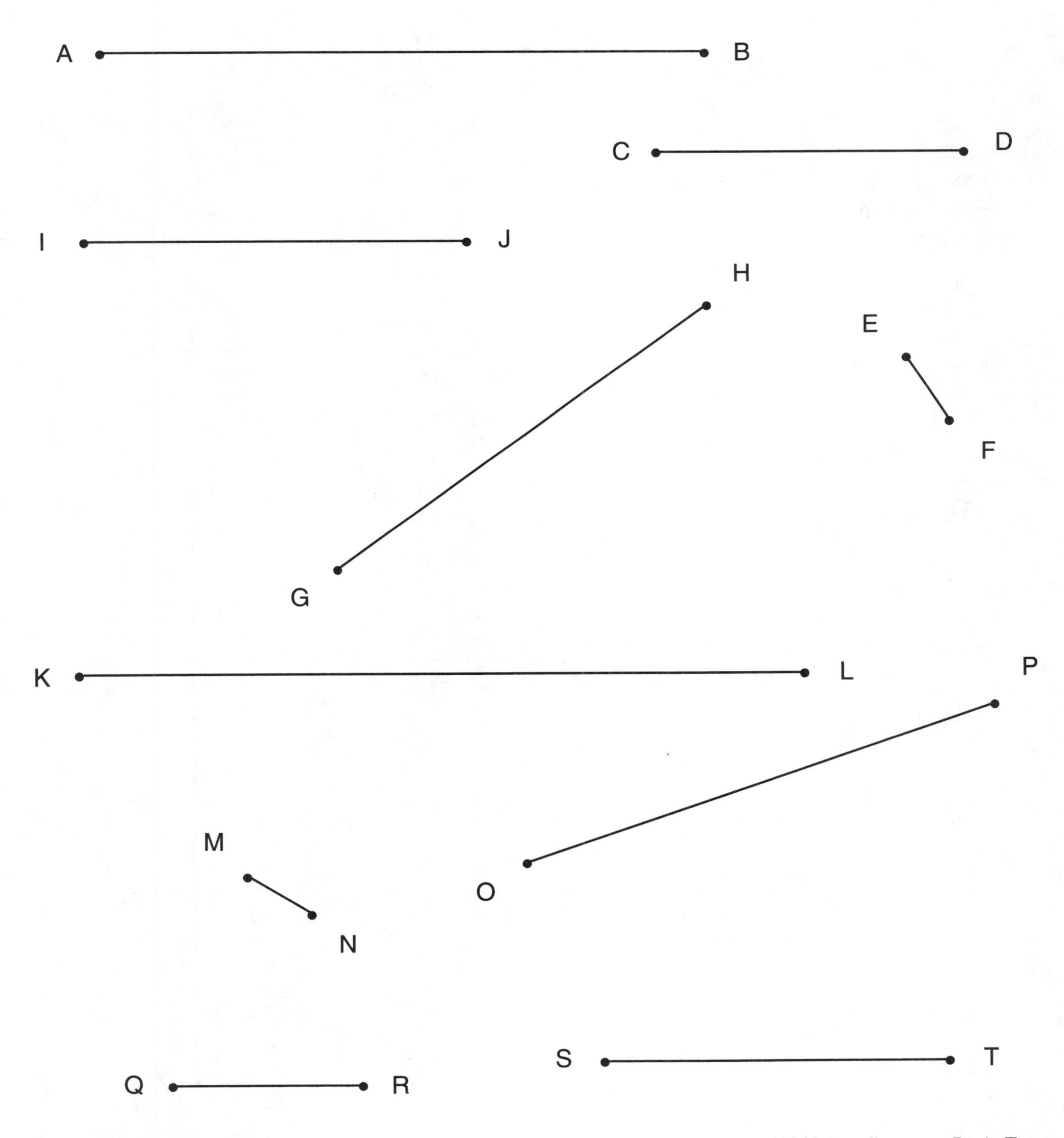

More Line Segments

Directions: Follow the steps on page 14 to copy the line segments below. If you do it correctly, you should get $\overline{BC} + \overline{CR} + \overline{RM} + \overline{MD} + \overline{DP} = \overline{EZ}$.

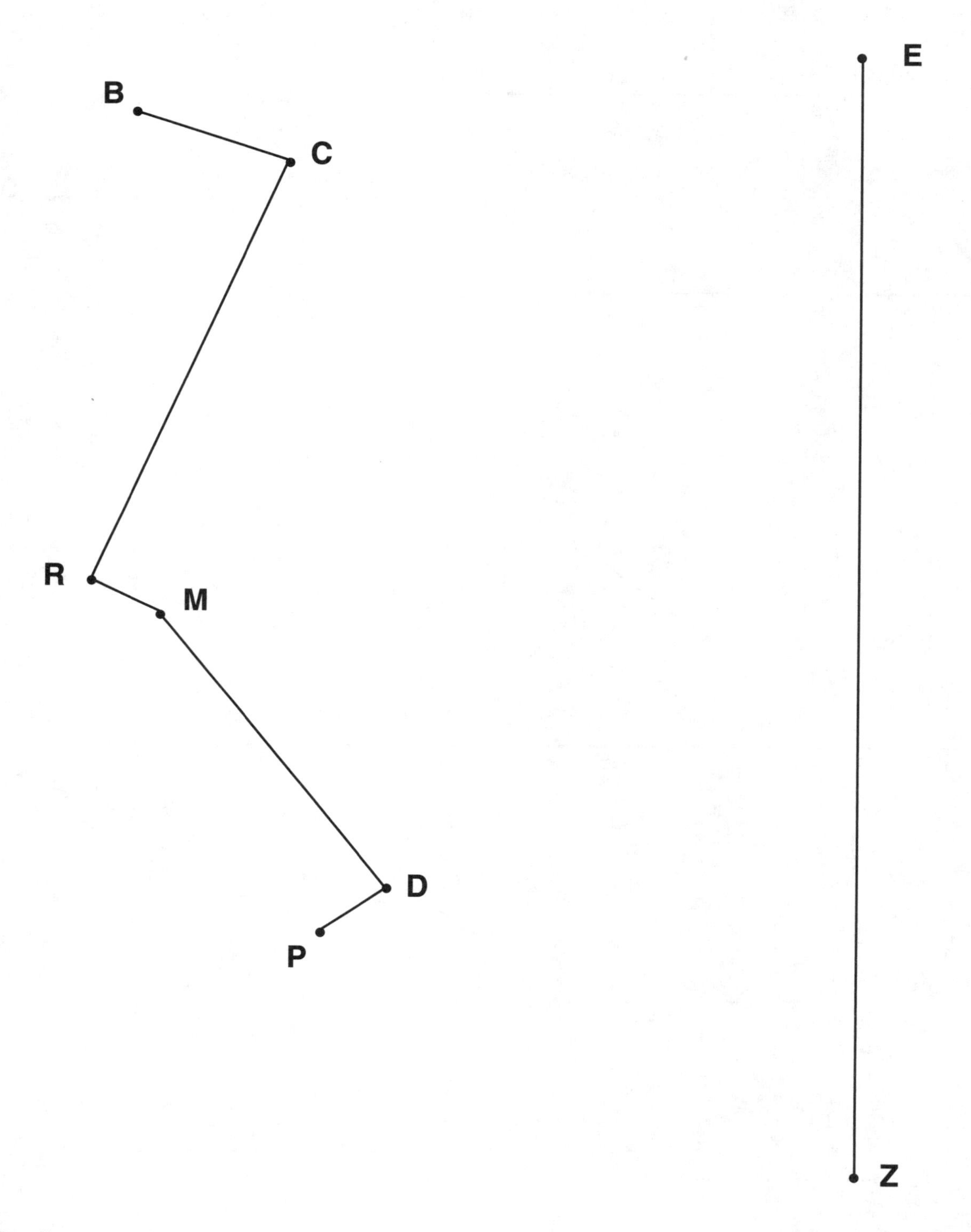

Simple Map Copying

Directions: For this activity, you will use a compass to enlarge a map to three times its size. Use the examples below to help you with this process. Measure a line on the original map with the width of your compass. Assign points A and B to the ends of the line segment. Place your compass point on point B and mark off the second segment. Your line is now doubled. Move the point of the compass to the pencil mark at the end of the second segment and measure the third length as you did the others. Use a straightedge to connect point A to the end of the line. Copy each line segment in the same way to complete your enlarged map.

Example

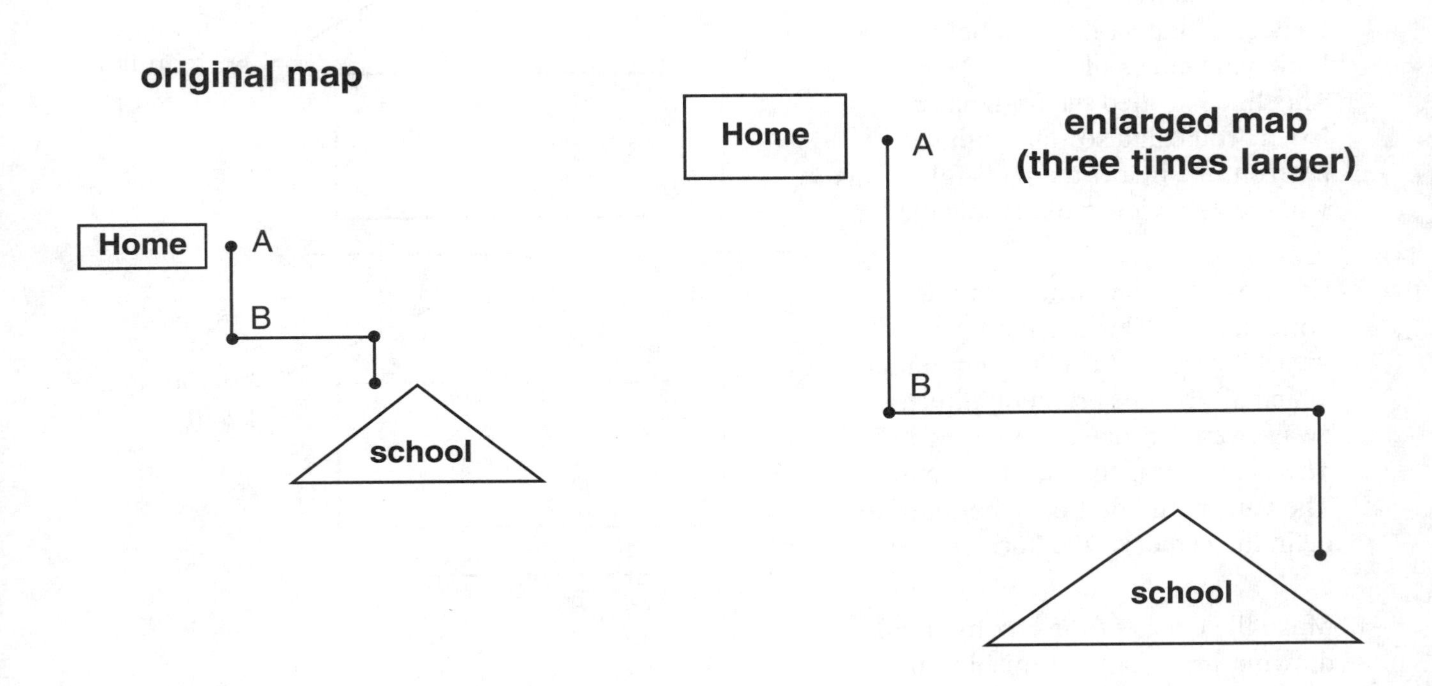

Use the method above to enlarge the following map on a 12" x 18" (30cm x 46 cm) piece of paper.

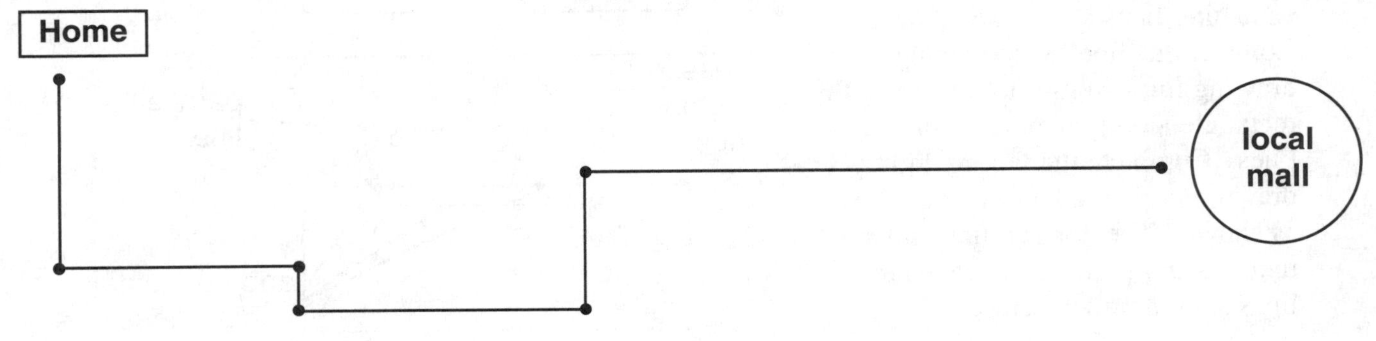

Perspective

Geometry can be used to solve many mathematics-related problems in real life. How does an artist, while working on a two-dimensional surface, solve the problem of creating a painting or drawing that has a three-dimensional look? This problem of perspective (a technique for making solid objects or spatial relationships drawn on a flat surface appear to have depth and look true to life) can be solved using lines in a specific way. The example below shows how to use lines to draw a perspective view of a rectangular solid, or box. Follow these basic steps to make a box. Create an interesting picture using several boxes. Make different shapes and sizes to add variety. You can include other designs or objects in your picture.

Directions

1. Draw a rectangle. Above the rectangle, draw a line parallel to the horizontal edges of your rectangle. This line is called the horizon line. Note: You can also start with the horizon line below the rectangle. This will create a view from below the box.

2. Draw a point somewhere on the horizon line. This is called the vanishing point. (This is the point at which all the lines running directly away from the viewer come together. The view seems to vanish.) **Note:** The vanishing point does not have to be in the center of the horizon line.

3. Make the edges of the box by lightly drawing lines, called vanishing lines, from the corners of the rectangle to the vanishing point.

4. Draw a parallel line across the vanishing lines above the rectangle. Connect this line to the rectangle by drawing lines from it to the top of the rectangle along the outer vanishing lines. Complete the box by lightly drawing horizontal and vertical lines as shown. The horizon line and any unnecessary parts of the vanishing lines should now be erased.

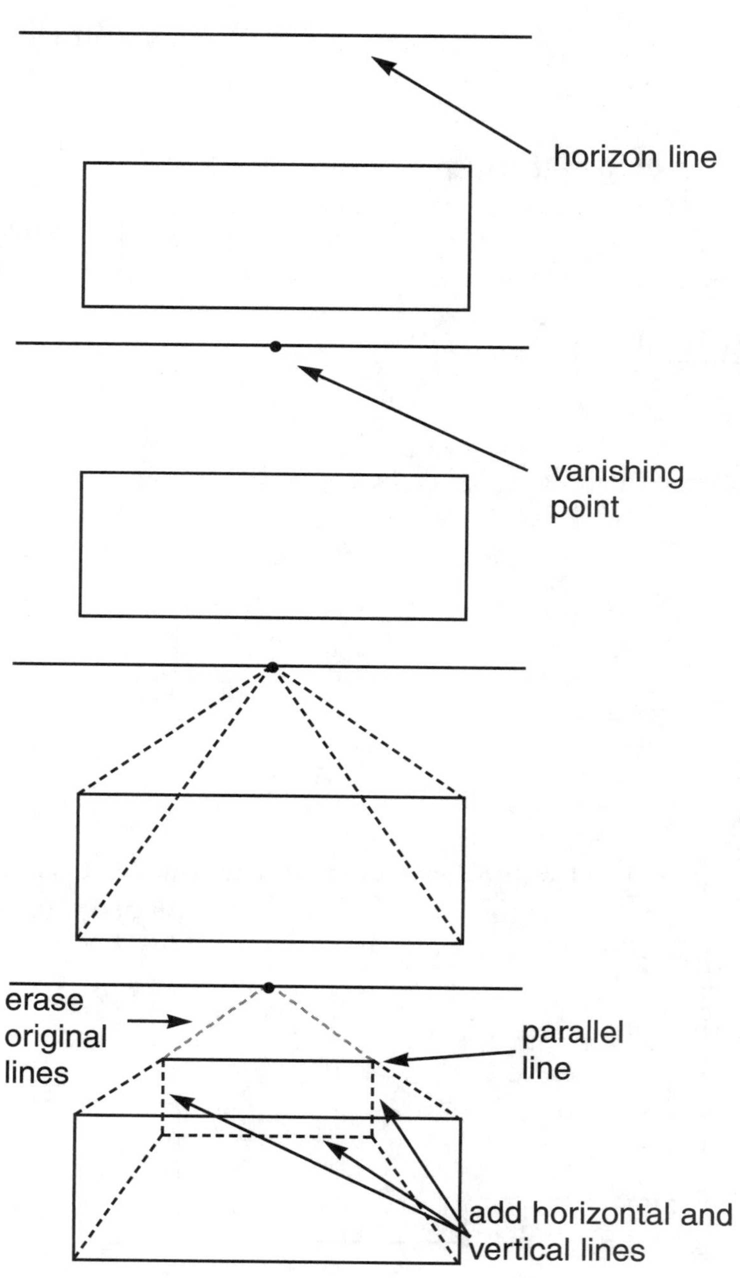

Perspective (cont.)

How does an artist create a perspective view of telephone poles, trees, or other objects disappearing from view in the distance? The steps below will show you how to create the vertical lines that will represent where you are to draw your objects. You will see how to create your own perspective view of a series of objects. When you are done, the objects will appear equally spaced as they move from large to small in the picture. Use this page to practice making a series of poles, trees, cornstalks in a field, etc. Then use page 20 to design a picture that includes one of these perspective drawings.

Directions

1. Make a horizon line and place a vanishing point (X) on it.
2. Draw a vertical line segment (AB) to represent the first of the series of objects (the one nearest to you as the viewer). Connect AB to the vanishing point.
3. Draw a second vertical line segment (CD) parallel to the first, at the place you would like to draw the second object. Make diagonal lines (BC and AD). Label the point at which they meet (Y).
4. Draw a line from X to line segment AB, passing through point Y. This line will become the center point for each of the objects.
5. To make the vertical line segment for the third object, draw a line from point A through the center point of line segment CD until it touches line BX. This now becomes the base of the next object.
6. Add the rest of the vertical line segments in the same way you made AB and CD. These line segments represent the positions for each of the objects you wish to draw.

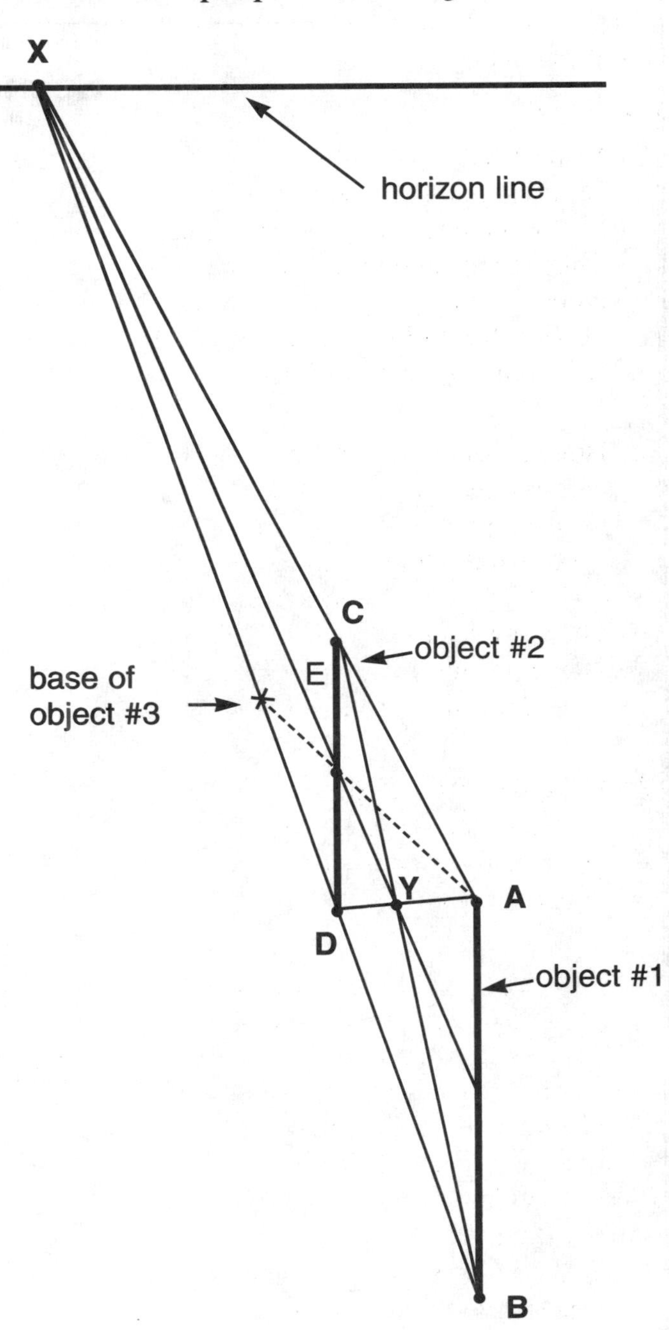

Perspective (cont.)

Directions: Use the information on page 19 to help you draw a scene that contains a series of objects moving toward a vanishing point. You could use such objects as rows of trees, telephone poles, or a vanishing road with road signs on each side, or a series of objects of your own choosing.

____

Scene Title

Pascal's Pattern

Seeing patterns and observing the relationship between numbers, lines, objects, etc., will help you understand geometry and its place in the world around you. In this exercise, you will look for patterns in sequences of numbers that have been arranged in a triangle, called Pascal's Triangle. This is a triangular array. The triangle was named after Blaise Pascal, a seventeenth century French mathematician who developed a triangular pattern of numbers arranged according to their properties.

Below is a portion of Pascal's Triangle. Look carefully at the relationship of the numbers and try to complete the next several rows of the triangle. When you are finished, write statements about some of the patterns you noticed.

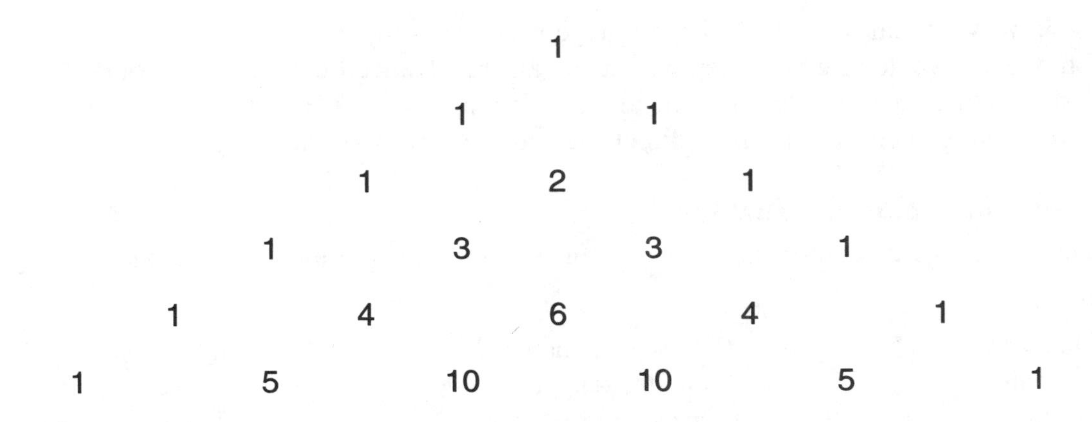

____

____

____

____

Challenge: Can you make a triangle of numbers that have interesting patterns in them similar to those in Pascal's Triangle?

Teacher Pages: Angles and Triangles

The following information corresponds to the exercise pages in this section. Where applicable, teacher information that may be helpful to the students' understanding of exercises is provided. For easy reference, the page numbers and titles of the exercises are given.

Page 25: Naming Angles

The challenge activity on this page gives students the opportunity to observe geometric figures carefully and identify shapes within shapes. Students should be able to demonstrate how they came up with their totals.

Page 28: Using Vectors

This exercise is very similar to LOGO, the language used for computer programming. It also introduces students to vectors, which they will see again in advanced mathematics courses. Vectors are used in industries where directional concepts are very important. Students are asked to create their own stories and maps using vectors as tools to direct the action or events of the story.

Pages 29 and 30: Triangle Puzzler

The information on page 29 can be used for reference as students complete the puzzler on page 30.

Solutions to page 29:

1. isosceles, acute
2. obtuse, scalene
3. right, scalene
4. isosceles, acute
5. acute, scalene
6. isosceles, obtuse
7. right, scalene
8. obtuse, scalene
9. obtuse, scalene

The challenge activity on page 30 encourages students to extend their knowledge of triangles by identifying types of triangles using specific criteria. In this case, each triangle must use at least one full side and at least one angle of the large triangle ABC.

Page 30 challenge solutions: AGC (obtuse, scalene); BAG (acute, scalene); BAQ (scalene, right); QCB (scalene, obtuse).

Page 31: Euclid's Thinking

The Greek mathematician, Euclid, described geometry as a system of reasoning. This activity familiarizes students with Euclid's axioms and postulates and challenges them to discover some of his theoroms. Depending on the students' knowledge of this topic, you may need to explain and demonstrate the following ideas to them: axioms are algebraic or arithmetical ideas that can be demonstrated but not proved; postulates are geometric ideas that can be proved. Axioms and postulates are used to create theorems (mathematical concepts that can be proved).

Page 32: Triangles

The angles of a triangle equal 180°. There is no way to prove this, but you can demonstrate this very easily. A straight angle is 180°. Take a large paper triangle and fold all the angles so that they equal 180°. After completing the activity, students are challenged to demonstrate some notions using triangles.

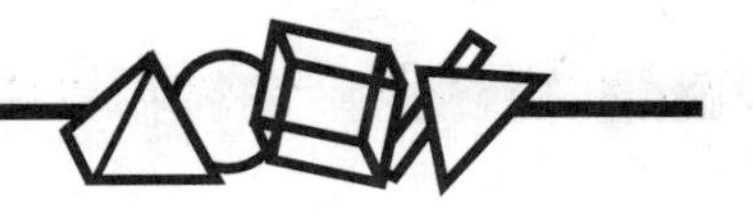

Teacher Pages: Angles and Triangles (cont.)

Page 33: Sides of a Triangle

The measure of one side is less than the sum of the measure of the others. A simple demonstration using cardboard strips and fasteners can also show this. In this activity, students are encouraged to apply Euclid's theorem using the information they have gathered about triangles.

Page 34: A Line and Point Postulate

After demonstrating this postulate, students are asked to make their own statements about lines, points, angles, or geometric shapes.

Page 35: Pythagoras and the Right Triangle

As they work through this activity, students learn that the right triangle has unique characteristics. Pythagoras first discovered this in the sixth century when he demonstrated the relationship between the two sides (legs) of the triangle and the hypotenuse. His theorem, $a^2 + b^2 = c^2$ is basic to the understanding of right triangles.

The most common right triangle is the 3-4-5 triangle, in which 3 and 4 represent the legs, and 5 represents the hypotenuse. This 3-4-5 combination is called a triple. There are 10 sets of triples based on the 3-4-5 right triangle that fall between 3 and 50. They are as follows:

3-4-5	9-12-15	15-20-25	21-28-35	27-36-45
6-8-10	12-16-20	18-24-30	24-32-40	30-40-50

Students are asked to use the Pythagorean Theorem to find as many of these triples as possible. In addition, encourage them to state any number relationships they find between and among the triples. **Note:** There are many more sets of triples between 3 and 100. If students wish to find some of these, they should be encouraged to do so. Below is a list of these other triples.

Other Triples based on set 3-4-5:	**Triples based on set 5-12-13:**	**Triples based on set 8-15-17:**	**Triples based on set 7-24-25:**	**Others:**
33-44-55	10-24-26	16-30-34	14-48-50	9-40-41
36-48-60	15-36-39	24-45-51	21-72-75	11-60-61
39-52-65	20-48-52	32-60-68		12-35-37
42-56-70	25-60-65	40-75-85		13-84-85
45-60-75	30-72-78			16-63-65
48-64-80	35-84-91			18-80-82
51-68-85				20-21-29
54-72-90				28-45-53
57-76-95				24-70-74
				33-56-65
				36-77-85
				39-80-89
				40-42-58
				48-55-73
				60-63-87
				65-72-97

Triangles from Points

The definition of a triangle is a figure with three sides, three angles, and three points. Triangles can be named by using three points in their correct sequence. For example, triangle ABC is a triangle with lines segments drawn from A to B, B to C, and C to A. Note the rule that the three points cannot be on the same line.

Assign letters to the points below and then connect the points to make as many triangles as possible. Using colored pens or pencils will help keep the drawings clear. When you have finished, complete the chart on page 25.

Naming Angles

On page 24, you used points to create many triangles. Each triangle you created has three angles that can be named in two ways. Here is an example.

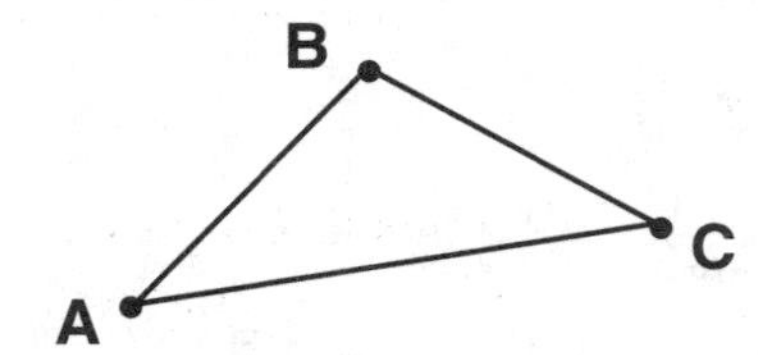

For this triangle, the angles can be named briefly as angle A, angle B, and angle C, or they can be named formally as angle ABC, angle BCA, and angle CAB. The symbol ∠ can be used in place of the word "angle."

Use the exercise on page 24 to complete the following chart. Name 10 triangles followed by the formal names for the three angles.

Triangle	First Angle	Second Angle	Third Angle

Challenge: Count the number of angles in this figure. Then, count how many triangles there are altogether. (Look for triangles within triangles also.)

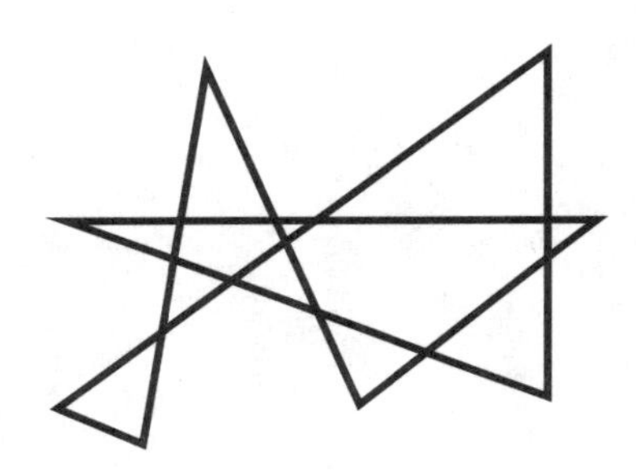

Angles

Each triangle has three angles. Angles have a vertex (point) and two sides (rays). Angles are measured by degrees. If one ray makes one whole revolution, it will sweep 360°

A right angle measures 90°. Any angle less than a right angle is an acute angle. Any angle greater than a right angle and less than a straight angle is called an obtuse angle. Now you have four angles you can refer to: *acute, right, obtuse*, and *straight*. (There is another angle not often used, which is an *oblique* angle. This is an angle that is more than 180° and less than 270°).

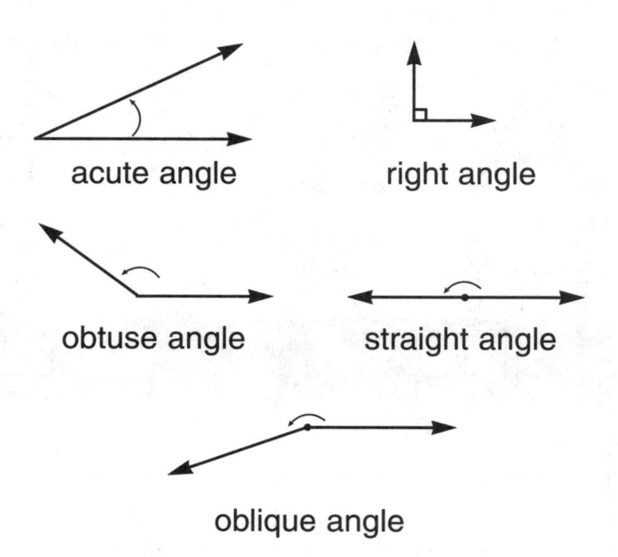

Find and name (acute, right, obtuse, straight, or oblique) all the angles in this figure. Make a chart showing all the angles you found and label each angle with its correct name.

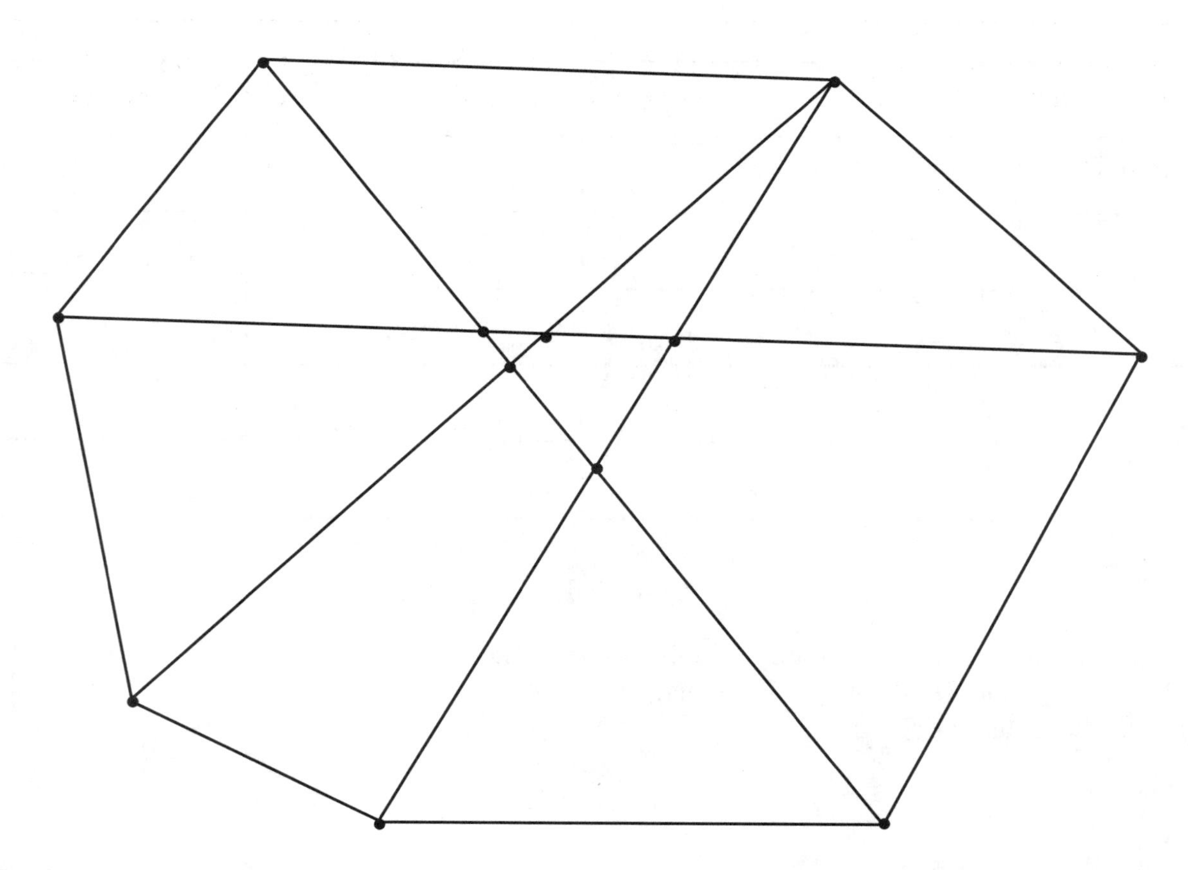

Alphabet Angles

Directions: Many capital letters are made entirely of line segments and angles. The angles for each of these letters are indicated below. On separate paper, make an enlarged set of these letters. (Your letters should be large enough for you to measure the angles.) Use a protractor to measure the angles. Then, in the box, write the total number of degrees for each letter.

A E F H I K L M

N T V W X Y Z

A = ______	E = ______	F = ______	H = ______	I = ______
K = ______	L = ______	M = ______	N = ______	T = ______
V = ______	W = ______	X = ______	Y = ______	Z = ______

Now create 10 words from these letters. Write the correct number of degrees for the inside angles of each letter in the word, add these, and write the total for each word. An example is provided.

Example: IT = 360° + 90° = 450°

1. ______________________________
2. ______________________________
3. ______________________________
4. ______________________________
5. ______________________________
6. ______________________________
7. ______________________________
8. ______________________________
9. ______________________________
10. ______________________________

Using Vectors

An angle defines the measurement between two rays that meet at one point. An angle also shows two directions. A *vector* is a ray that shows direction as well as distance. The sample below shows the path of a bear. Find out how far away and in what direction the bear wandered from the point at which it began. Use a protractor and a ruler to follow the vectors described in steps 1–4. For this activity, 0° represents north and one inch (2.54 cm) is equal to one mile (1.6 km). Set up your protractor so that 0° is north, 90° is east, 180° is south, and 270° is west. Position the protractor this way for each vector.

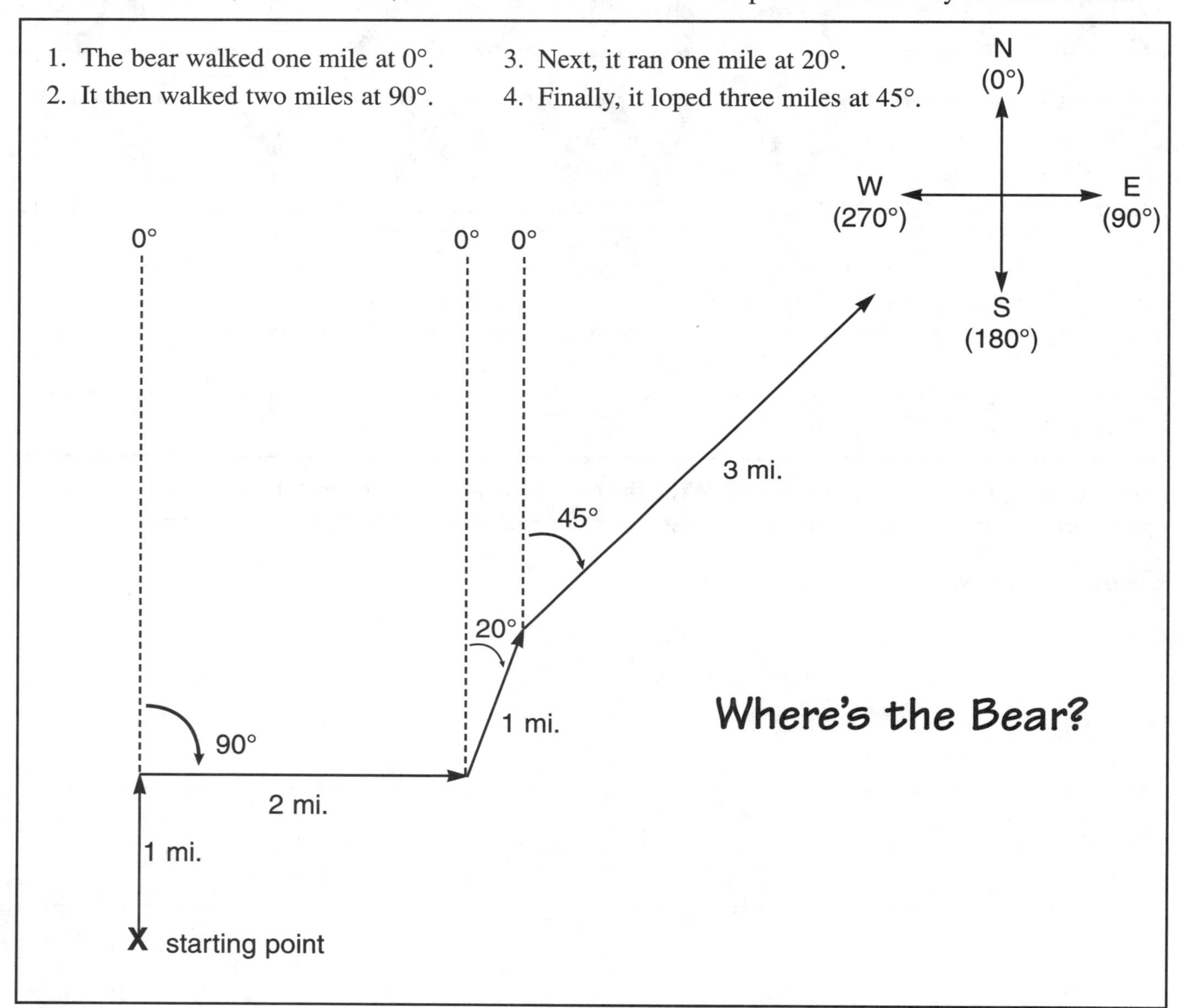

Create a story in which the character(s) move from one place to another. Be sure to draw a map beforehand. Then give your story to another person. If that person follows the story but doesn't end up with the same map as yours, find out why. The reason usually is that either the directions were not followed correctly or the directions were unclear. In any case, try to correct the situation so that the story and the map can be followed.

Triangle Puzzler

Kinds of Triangles

Triangles are named by the types of angles and sides they contain.

Directions: Read the chart below to find out the names of triangles and their descriptions. Identify the types of triangles at the bottom of the page. Note: Some of the triangles can be identified in more than one way. Explain all possibilities. Then, use what you have learned to solve the puzzler on page 30.

Type of Triangle	Description
equilateral	three sides of equal length and three angles of equal measure
isosceles	at least two sides must have the same measure
scalene	no equal sides
right	one right angle (90°)
acute	all angles less than 90°
obtuse	one angle greater than 90°

1.

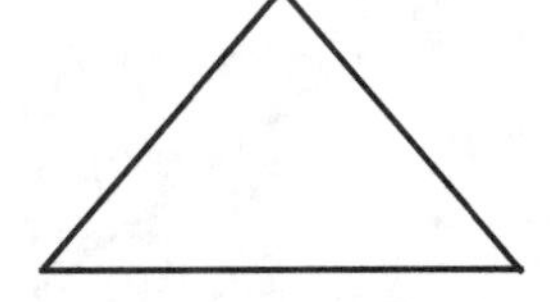

2.

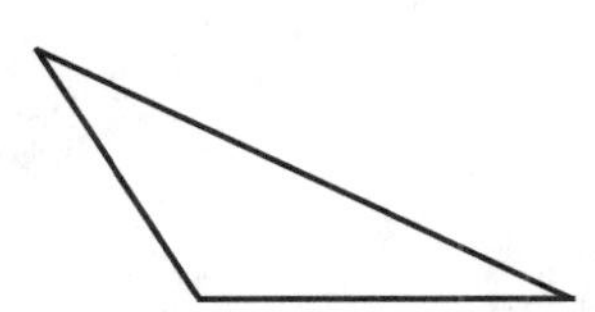

3.

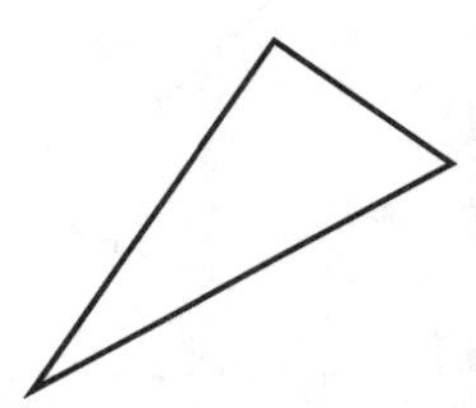

4.

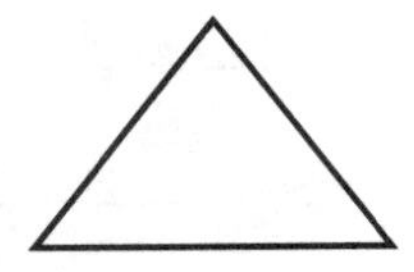

5.

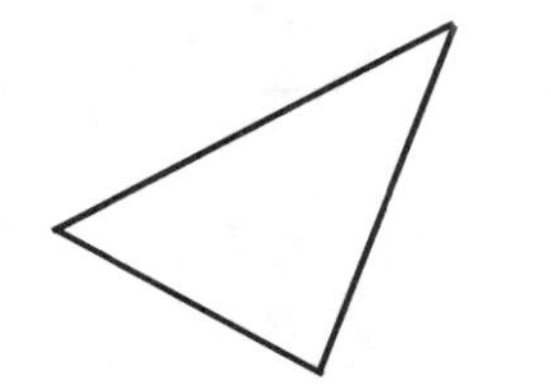

6.

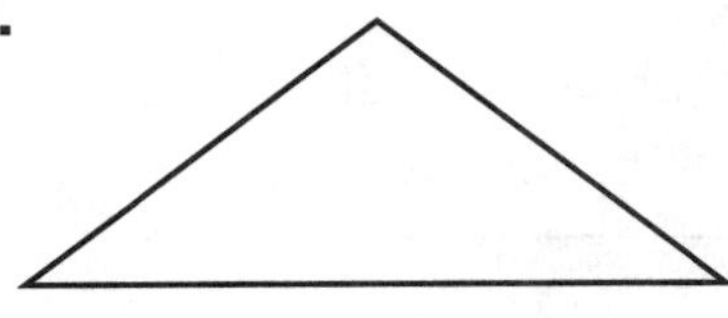

7.

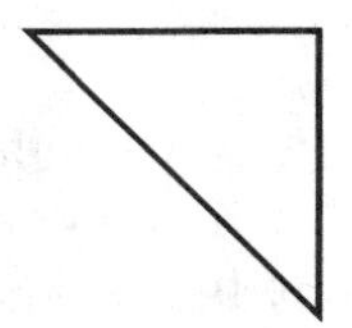

8.

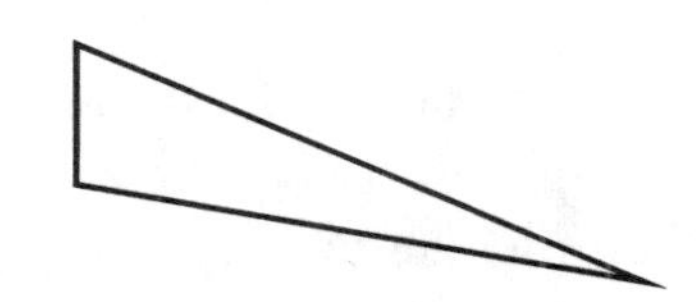

9.

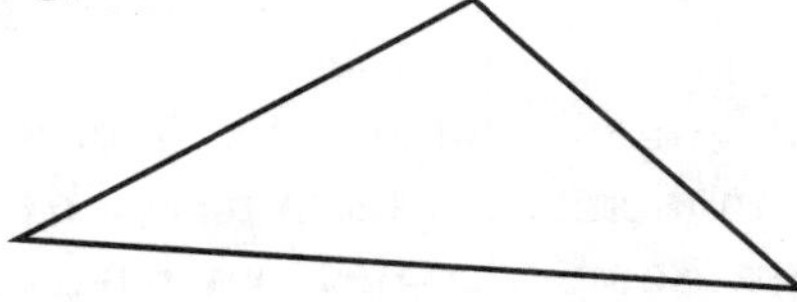

Triangle Puzzler (cont.)

Directions: Use the information on page 29 to find as many kinds of triangles as you can in the figure below. List them on the lines.

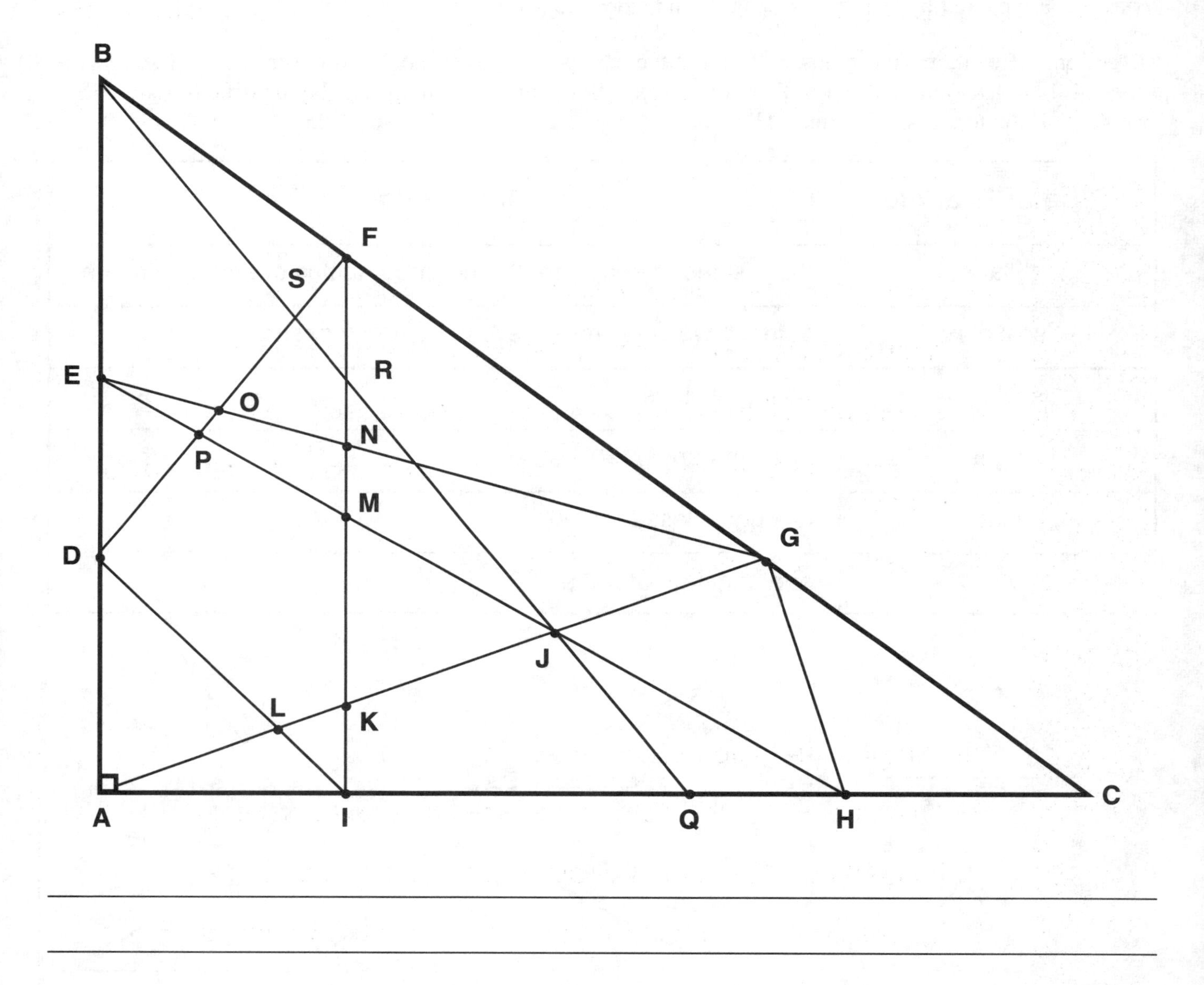

____

____

____

____

Challenge: Find all the triangles you can that have at least one side and one angle of triangle ABC within them. List them on the back of this paper. Label the kind of triangle for each one you find. Note: Some triangles may have more than one name. For example, a triangle may be scalene and obtuse.

Euclid's Thinking

The formulation of the branch of mathematics known as geometry is attributed to Euclid, who lived around 300 B.C. Euclid is best known for his 13 books, known as *Elements,* which contained over 200 postulates (assumptions), five "common" notions (axioms), and 465 theorems. In prior years, students who studied geometry were made to memorize Euclid's book in its entirety.

Those who study geometry can use these axioms and postulates to come up with new statements (theorems).

Postulates

Here are examples of four well-known postulates:

1. A straight line can be drawn between any two points.
2. Any straight line segment can be extended infinitely.
3. A circle with any radius may be described around a point as a center.
4. All right angles are equal to each other.

Notions (Axioms)

The five notions (axioms) are as follows:

1. Things equal to the same thing are equal to each other.
2. If equals are added to equals, the sums are equal.
3. If equals are subtracted from equals, the differences are equal.
4. Things which coincide with one another are equal.
5. The whole is greater than any of its parts.

There are too many theorems to include here, but they can be found in most advanced geometry textbooks for any student who is interested in reading them.

Triangles

Directions: Discover one of Euclid's theorems for yourself. Measure the angles in each of the triangles below. Find the total degrees for each of the angles. What did you discover? Write your findings in a statement that expresses one of Euclid's theorems.

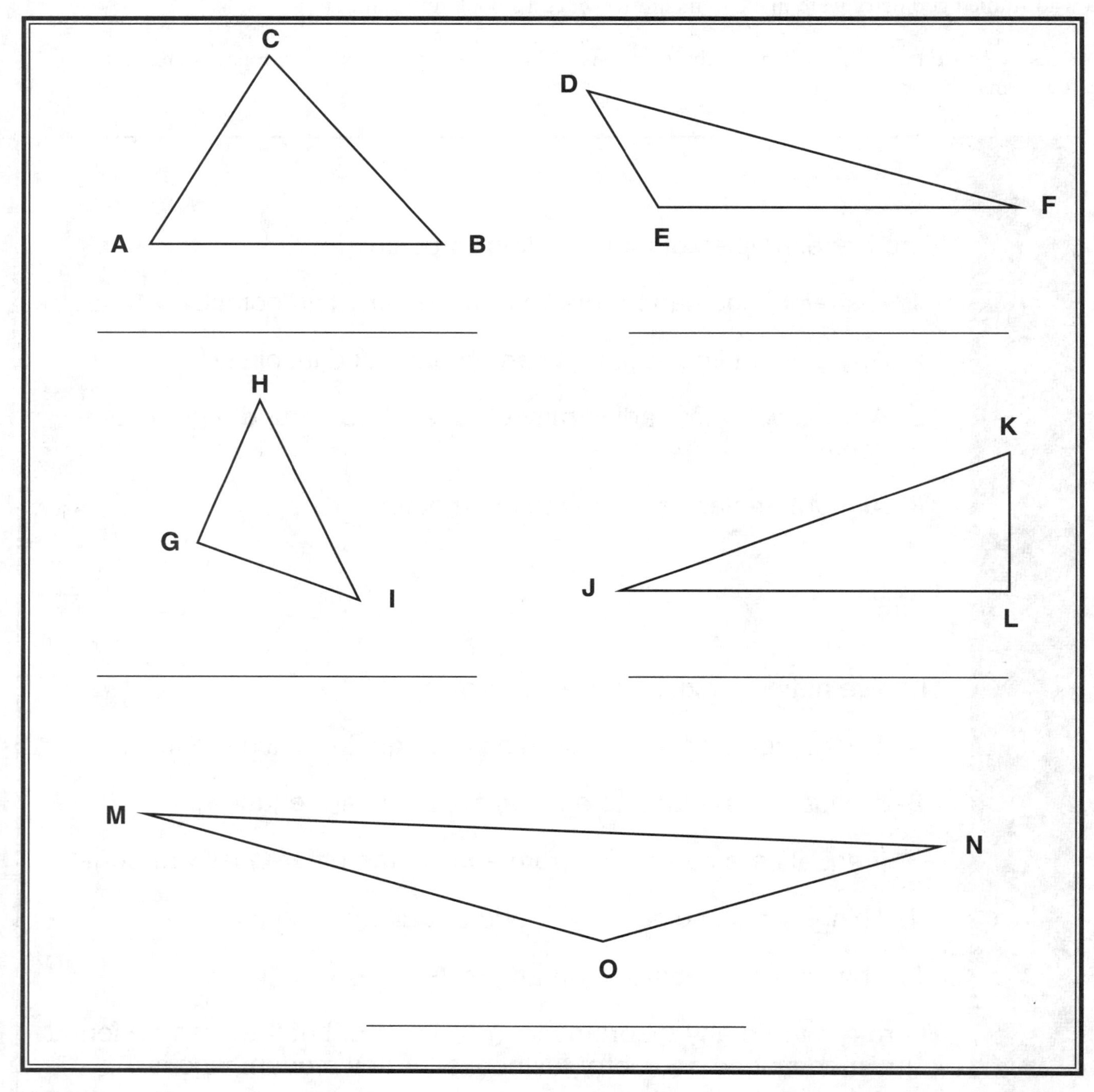

Challenge: Choose a few of the five notions given on page 31. Give examples of each using what you know about triangles. State which notions you are demonstrating.

Sides of a Triangle

Directons: Measure the sides of these triangles (in centimeters) to see if you can discover another of Euclid's theorems. Write it on the lines provided.

1.

C

A B

side AB ____ side AC ____ side CB ____

2.

F

D E

side DE ____ side DF ____ side FE ____

3.

H

I

G

side GI ____ side GH ____ side HI ____

4.

L

J K

side JK ____ side JL ____ side LK ____

5.

O

M N

side MN ____ side MO ____ side ON ____

6.

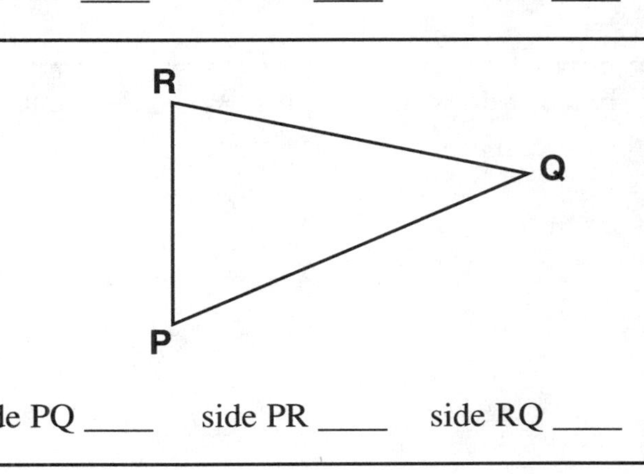

side PQ ____ side PR ____ side RQ ____

____

____

____

____

A Line and Point Postulate

A postulate is a statement that cannot be proved but can be demonstrated. In the following example, you will learn more about postulates.

Directions: Draw a line. Make a point outside the line. Draw as many lines through the point as you can (5 or more). You will note that there is only one line that can be drawn through the point parallel to the line.

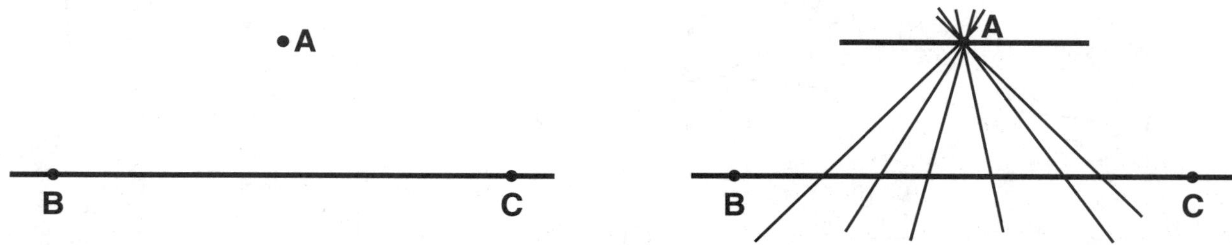

You can make this finding into a statement, or postulate: "Although many lines can be drawn through a point outside a given line, only one parallel line can be drawn."

When you draw a line not parallel to the given line, you create an angle. If you do this with two lines, you can make a triangle. If you connect a point outside a line segment with other points on the line segment, you create other triangles.

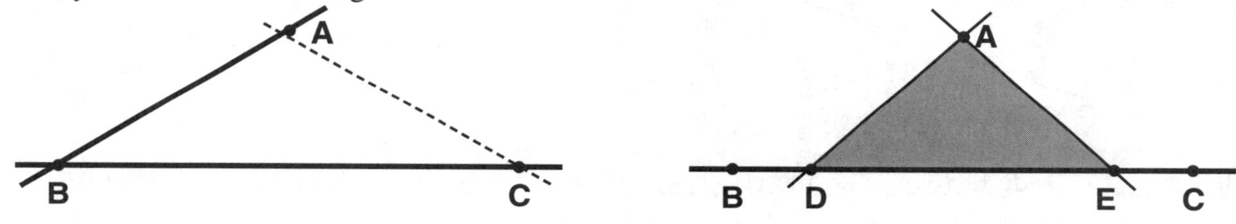

Practice drawing triangles. Write the name of the triangle and name the parts.

Example:

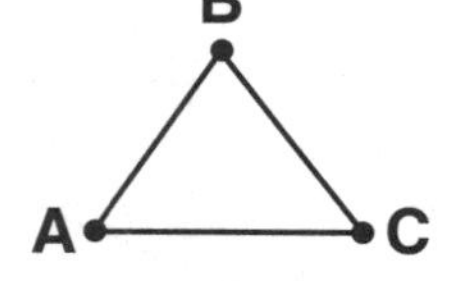

$\triangle$ A B C

$\angle$ A, $\angle$ B, $\angle$ C

sides $\overline{AB}$, $\overline{BC}$, $\overline{CA}$

Challenge: Think about statements or assumptions you can make about lines, points, angles, or various geometric shapes. Make drawings that demonstrate your statements.

Pythagoras and the Right Triangle

Pythagoras was a Greek mathematician and philosopher, famous for his Pythagorean theorem. This theorem states that in a right triangle, the square of the length of the hypotenuse is equal to the sum of the squares of the lengths of the other two sides. The hypotenuse is the side opposite the right angle.

The Pythagorean theorem can be written as the formula $a^2 + b^2 = c^2$, where c is the hypotenuse, and a and b are the lengths of the other two sides. (See illustration.)

The right triangle that we most commonly associate with the Pythagorean Theorem is the 3-4-5 triangle in which 3 and 4 represent sides a and b, and 5 represents the hypotenuse.

This is known as a Pythagorean Triple. There are many more sets of Pythagorean Triples that can be found using the formula $a^2 + b^2 = c^2$. In fact, there are 50 sets of numbers (triples) in which each leg and hypotenuse is less than 100!

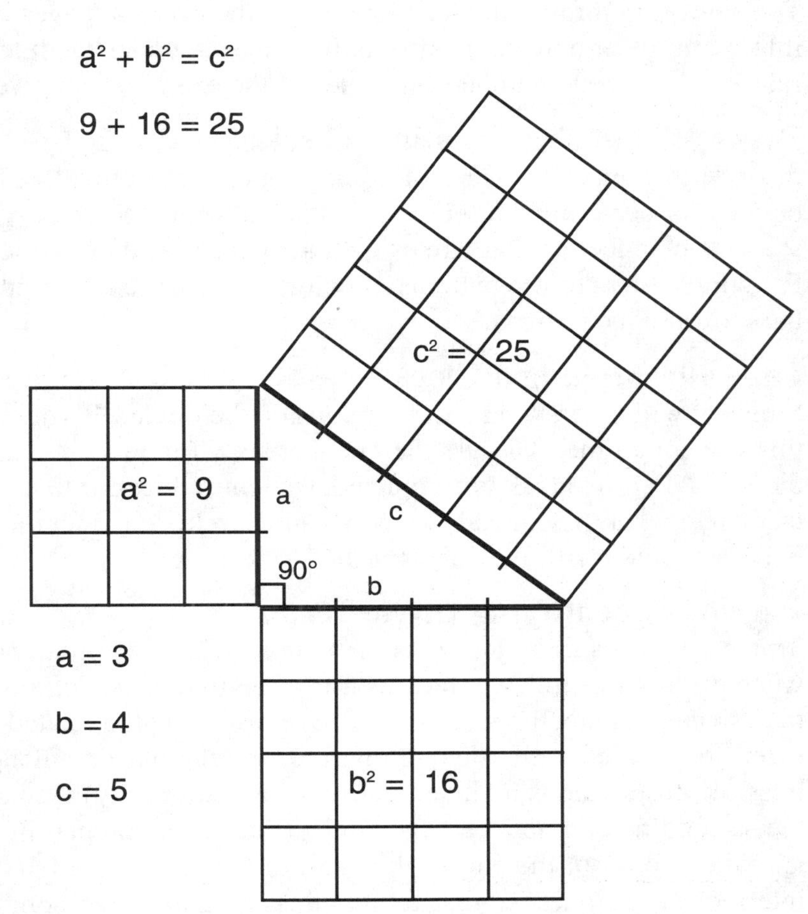

Directions: There are ten sets of triples (positive whole numbers) in which each leg and hypotenuse is less than 50. Can you find them? Use the formula above to help you. Write the triples in the box below. When you have discovered some or all of these, see if you can find any number relationship among the triples.

Teacher Pages: Circles

The following information corresponds to the exercise pages in this section. Where applicable, teacher information that may be helpful to the students' understanding of the exercises is provided. For easy reference, the page numbers and titles of the exercises are given.

Pages 38 and 39: Creating Circles

A circle is defined as the set of points equidistant from a fixed point. Students are asked to define a circle after performing a set of steps that show how to create a circle. After completing this activity, students practice making circles with a compass as they create art designs in the frame on page 39. Encourage them to use patterns of color, to use large and small circles, and to intersect circles to create the most unique designs.

Page 40: Making a Clock

Students will be drawing a hexagon inside the circle. If your students need a review in clock reading, this is a good time. Discuss the fact that sixty (minutes in a hour) is a factor of 360 (degrees in a circle). After the clock is completed, students are asked to use the steps they followed in creating a clock and what they already know about a circle, to explain how a clock and a circle are related and why the circle forms the basis for the clock design.

Page 41: Identifying Circle Terms

This page serves as an introduction and/or review of important terms used when discussing circles. Students are asked to use the definitions as a guide to drawing and labelling a circle. They must show the related terms. As an extension, students are asked to observe the relationship among these parts. Possible observations include the following: (similarity) the tangent line and the secant line both intersect the circle, (difference) no part of the tangent line is in the interior of the circle while part of the secant line falls within the interior of the circle; (similarity) a radius and a diameter both touch the center of a circle, (difference) a radius is one-half the length of a diameter and it only touches the circle at one point.

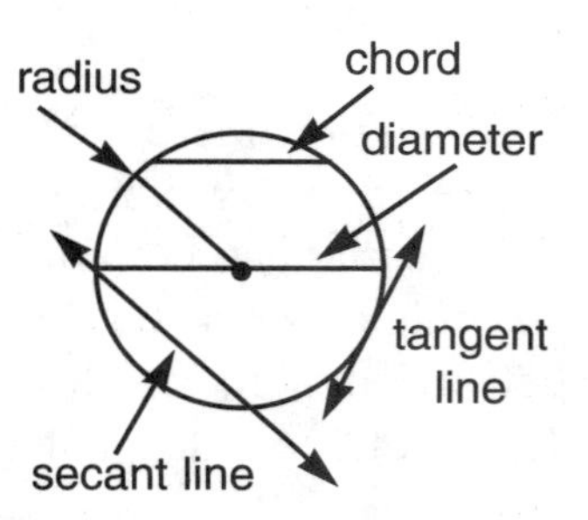

Page 42: Squaring Off a Circle

After students learn to make squares in circles, they are asked to make a series of squares whose corner points are equidistant along the circumference. One way to accomplish this is to first divide 360° by the total number of corners (points on the circumference of the circle) desired. For example, if you wish to make three squares (12 points/corners) divide 360° by 12. Use this quotient (in this case, 30°) to make interior angles from the radius. The points where the radii intersect the circle can then be used to make the squares. The resulting design is very interesting. Encourage students to color or shade the patterns to create unique effects.

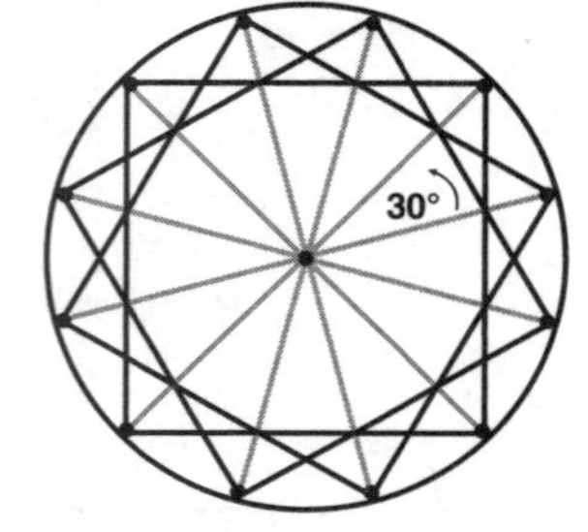

Example:

3-square pattern

12 points (4 corners x 3 squares)

$360° \div 12 = 30°$

Teacher Pages: Circles (cont.)

Pages 44 and 45: Types of Circles

Students use the information about concentric, circumscribed, internally tangent, externally tangent, inscribed, and eccentric circles, as well as inscribed and circumscribed polygons (page 44) to create a picture in the frame on page 45. (See illustration.)

Sample picture for page 45 (using a variety of circles)

Page 46: Pi and Circumference

Students are asked to calculate the value of pi by first measuring the circumference and diameter of an object and then dividing the circumference by the diameter. After completing the chart, students arrive at a rough calculation of the value of pi. Students must then complete the problems at the bottom of the page to find the actual value of pi. Have students try the challenge activity to extend their understanding of the relationships of the values of pi, the circumference, the radius, and the diameter.

Solutions: The value of π in problems 1–4 is 3.14. *Challenge:* a. $\pi\ ^{C}/_{d}$ b. $\pi = (^{C}/_{2}) \div r$

Page 47: Area of a Sector

In this activity, students work from the whole of an object to its parts. They use what they know about the area of a circle, the sector angle, and the radius to find the area of the sector. The example shows that once the sector is interpreted as a portion (fraction or percent) of the circle and the area of the circle is computed, the area of the sector can be found by multiplying the two.

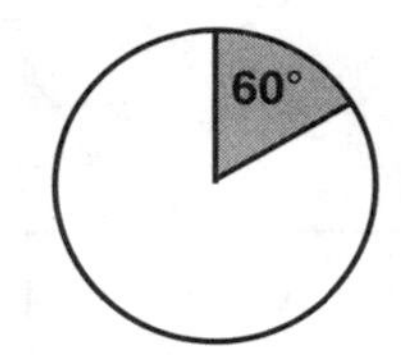

sector angle represented as a fraction or percent $\left(\frac{n}{360^\circ}\right)$ x area of the circle (πr^2)

Answers to page 47:

1. 1.413 cm^2
2. 5.024 cm^2
3. 19.625 cm^2
4. 219.8 cm^2
5. 235.5 cm^2
6. 2.826 cm^2

Page 48: Finding Areas Within Areas

Students will need to know the formulas for finding the area of a triangle and a circle.

In the first diagram, they can find the area of the large circle, subtract the area of the square (four right triangles), and add the area of the smaller circle. The shaded area is 21.38 cm^2. (area of large circle = 50.24 cm^2; area of square = 32 cm^2; area of small circle = 3.14 cm^2) More advanced students can also use the formula for the area of a segment $\left[\left(\frac{a}{360^\circ}\right) \cdot \pi r^2 - \frac{1}{2} bh\right]$.

In the second diagram, the outermost shaded area can be found by subtacting the area of the second largest circle from the largest circle. Do the same for the two smallest circles to find the smaller shaded area. Then, find the sum of the shaded areas. This activity can also be seen as a problem in finding the area between concentric circles. The formula is $A = (R^2 - r^2)\pi$, where the R represents the radius of the larger circle and r represents the radius of the smaller circle. To find the area between two concentric circles, simply square the larger radius, square the smaller radius, find the difference, and multiply the difference by π. Challenge students to try their own methods and compare them to this method.

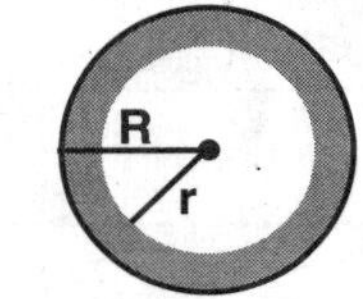

$A = (R^2 - r^2)\pi$
R = radius of larger circle
r = radius of smaller circle

Solution: smaller shaded area = 9.42 cm^2, large shaded area = 21.98 cm^2, total = 31.4 cm^2

Creating Circles

What is a circle? Use a piece of paper, a piece of string, a pencil, and the following directions to help you write a definition for a circle. When you have completed this activity, create a unique circle design in the frame on page 39.

Directions

1. Make a point in the center of the paper. Label it point P.
2. Cut a piece of string no longer than the distance from the center dot to an edge of the frame.
3. Make loops at the ends. One loop should be large enough to slide a pencil through. Your index finger should fit comfortably through the other loop.
4. Hold the index finger loop end at point P. Place the pencil through the other loop.
5. Hold your index finger firmly on point P. In a clockwise motion, begin making closely-spaced dots with the pencil. Make sure the string is taut. This may take a little time. (Remember that in a circle the line appears solid because the points are much closer together than you could possibly may them.)

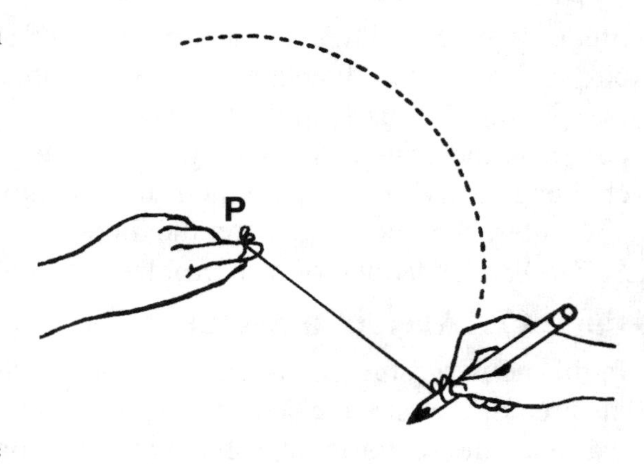

When you are finished, use the information above to write your own definition of a circle.

Circle

Definition: ____

____

____

____

____

____

____

____

____

Creating Circles (cont.)

Directions: Use a compass to create a unique design in the frame below. Add color or a variety of design patterns to your work of art.

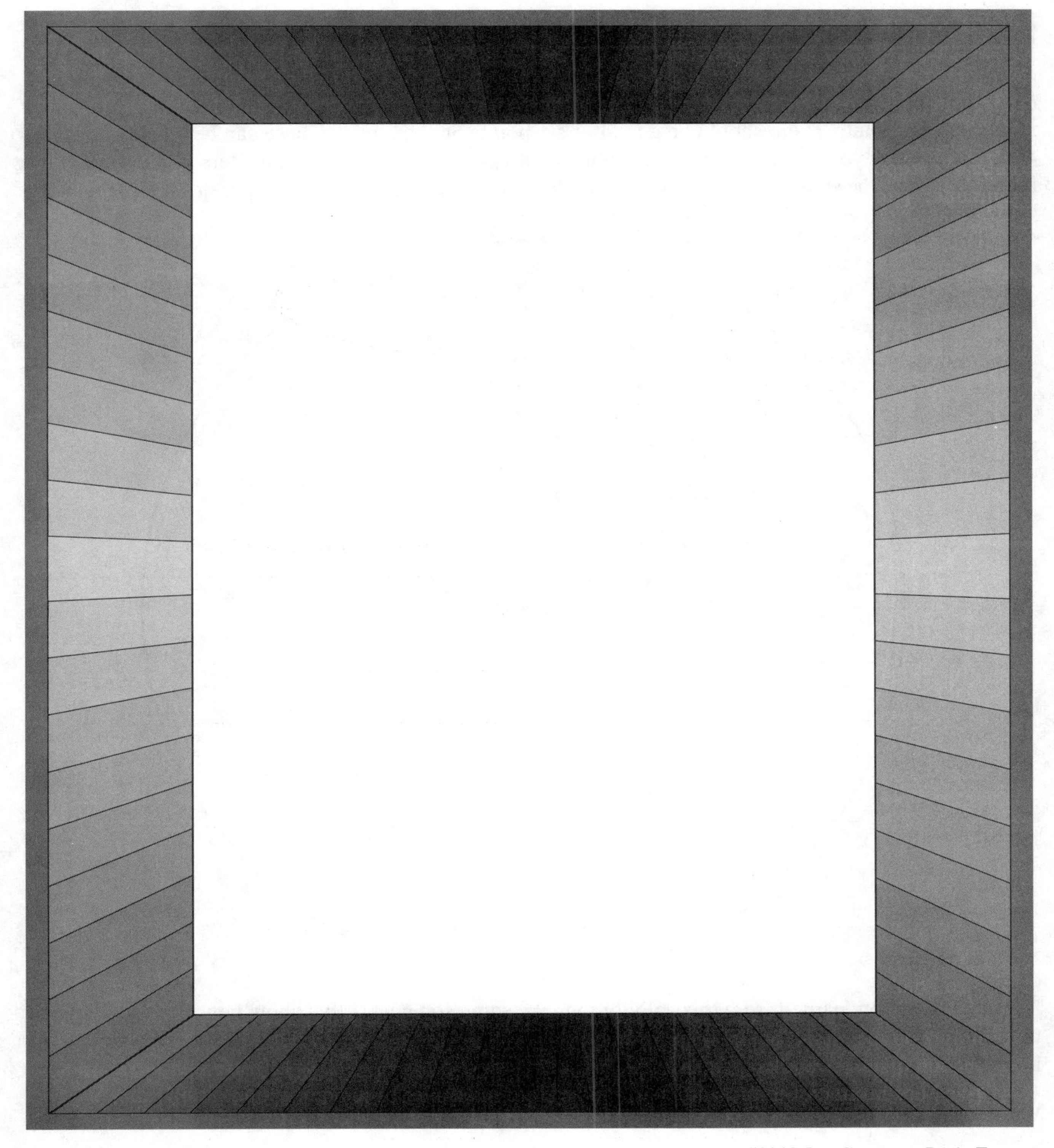

Making a Clock

In this activity you will discover basic concepts about circles and use them to make your own clock. Follow the directions and use the circle below to create your clock. You will need a compass.

Directions: Open the compass to match the length of the radius. At the top of the circle, make a point. Place the sharp tip of the compass on this point. Mark an arc, equal to the length of the radius, on the circle. Move the sharp tip of the compass to the point where the arc you have drawn intersects the outside of the circle, and make another arc. Continue to make arcs around the circle until you are back at the starting point. There should be a total of six points on the circle. These can be labeled as 12, 2, 4, 6, 8 and 10 on your clock. Decide how you can accurately place the odd numbers 1, 3, 5, 7, 9, 11 in between. Mark these on the clock. Give your clock hands, and decorate the clock in a unique way.

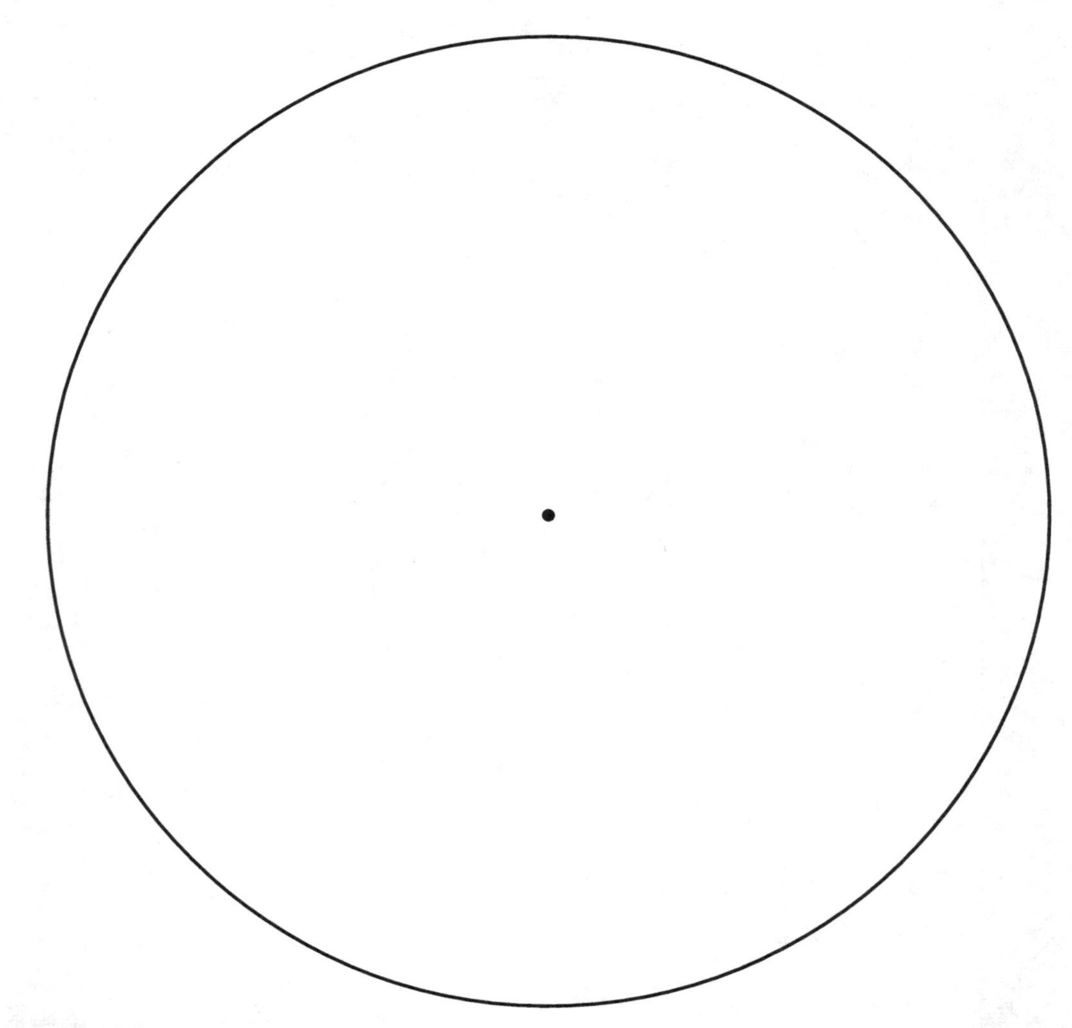

Now use what you learned from this activity to make some statements that show how a clock is related to a circle. Here are two facts to help you get started: A circle has 360°; There are 60 minutes in an hour and 60 seconds in a minute.

Write your ideas on the back of this paper.

Identifying Circle Terms

Directions: A circle is a set of points that are the same distance from a given point. Use a compass to draw a circle inside the box below. Next, read each of the following terms and draw a sample of each.

1. **radius**—line segment from the center of a circle to a point on the circle
2. **chord**—line segment that connects any two points on a circle
3. **diameter**—chord connecting the center of a circle to any two points on that circle
4. **secant line**—straight line that passes through any two points on a circle
5. **tangent line**—straight line that passes through only one point on a circle

Challenge: Look at the illustration you just completed. Think about how some parts of the circle are the same and how some are different. Write your observations on the back of this paper.

Squaring Off a Circle

Directions: Use the following steps to draw a square inside the circle below. Figure out how to add at least four more squares in the same way, so that all corners (points along the circle) are the same distance apart. (Hint: Remember that there are 360° in a circle.) When you have completed the squares, your design should be symmetrical.

Steps

1. Use your straightedge to lightly draw a diameter. Label it AC.
2. Use your protractor to measure a 90° angle from the center of the circle, and lightly draw a second diameter perpendicular to the first one. Label this diameter BD. You should have four points on the circle, points A, B, C, and D.
3. Use your straightedge to connect the four points to create square ABCD.

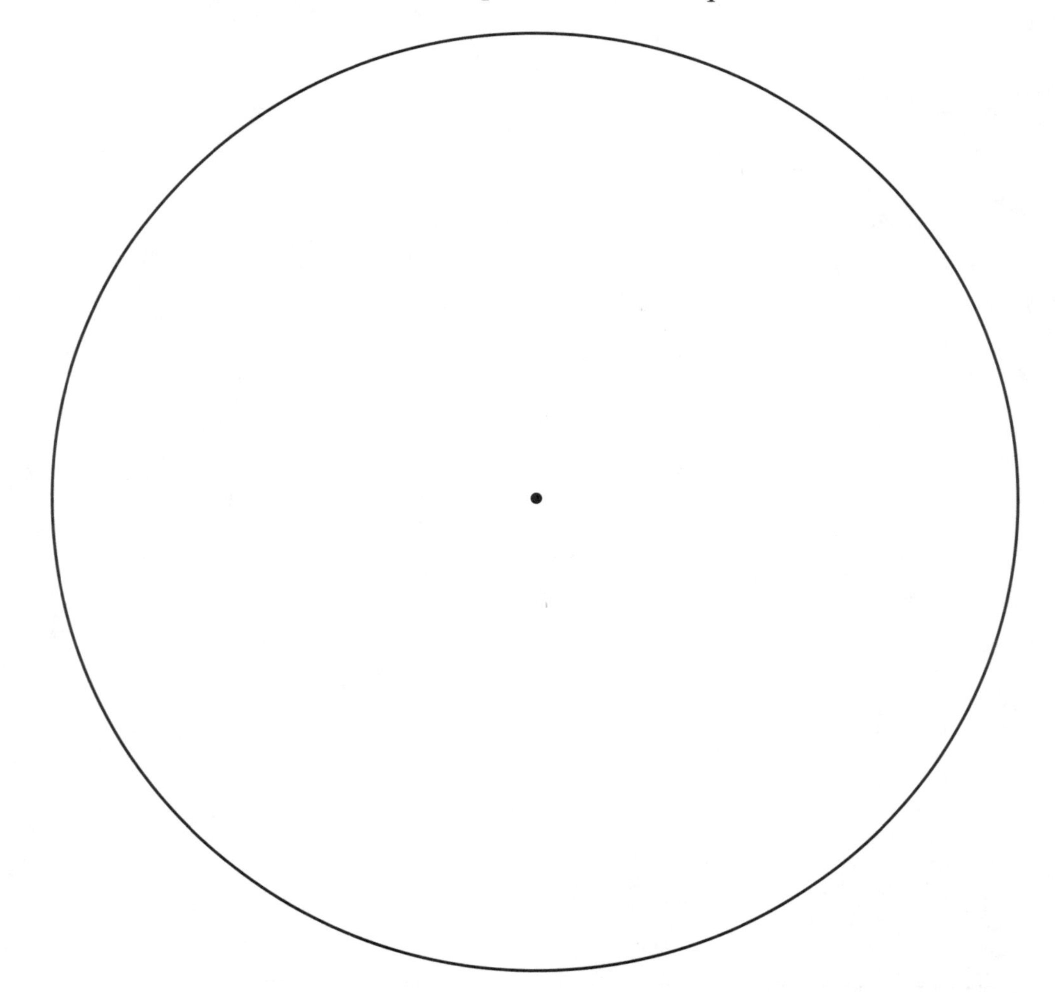

Challenge: Create a string art design using this method and the following materials: straightedge, compass, cardboard, hole punch, and yarn or thread of different colors.

Patterns Using a Circle

Directions: Make six equidistant points on the circle as you did on page 40. From each point, make a circle with a radius equal to the beginning circle. Note that the pattern you have created in the first circle resembles a flower. Continue drawing circles at all the points of intersection. Stop when the circle patterns go outside the box. How many points of intersection did you make? Can you make a pattern of circles starting with more than six equidistant points? What would the pattern look like? Create other patterns on larger pieces of paper. Color the patterns to create an interesting design.

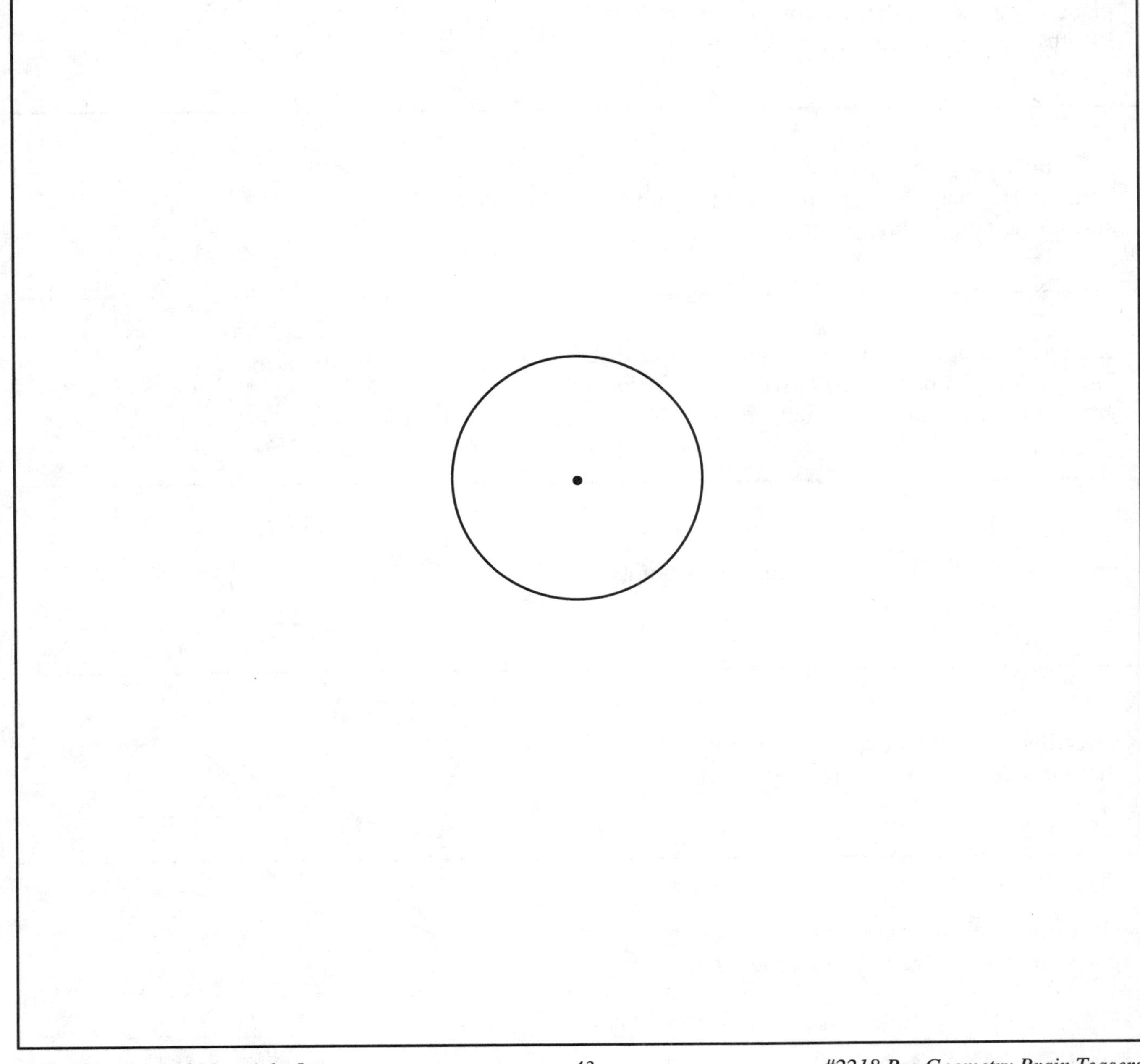

Types of Circles

The definition of any circle is a set of points, all of which are the same distance from a center point. There are specific names for circles as they relate to each other or with polygons. Read the following definitions and look at each illustration. Then, use the information on this page to complete the activity on page 45.

Type/Description	Illustration
concentric—two or more circles that occur in a plane, with each of the radii measuring different lengths and sharing the same center point	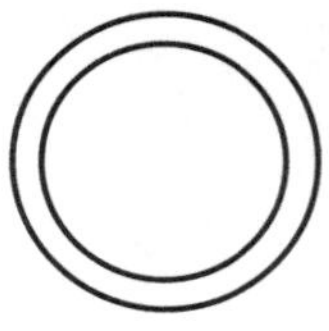
internally tangent—tangent circles in which both circles share the same point on the same side of the tangent line to the point	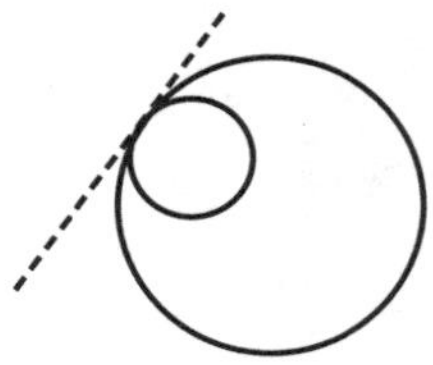
externally tangent—tangent circles (circles that intersect only at one point) in which the circles are on opposite sides of the tangent line	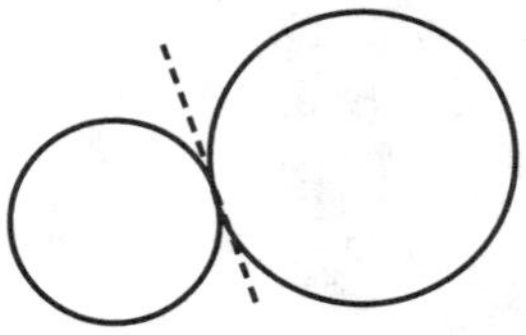
eccentric— circles with different center points	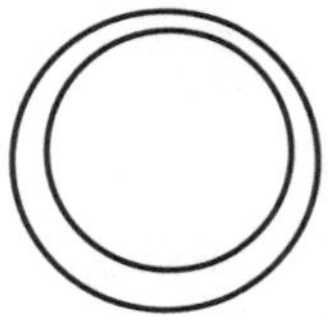
inscribed—circles situated within polygons so that the sides of the polygon are tangents	
circumscribed—circles with vertices of a polygon meeting the circumference	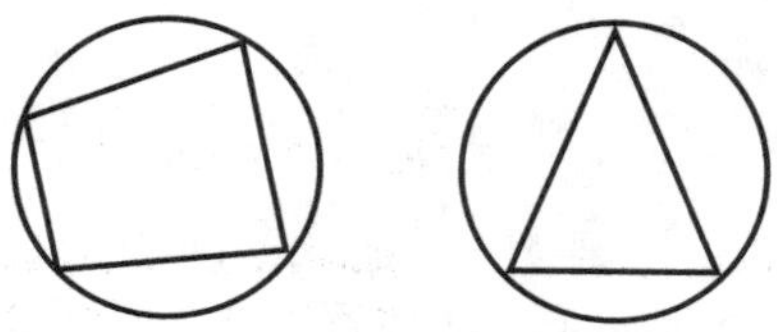

Types of Circles (cont.)

Directions: Create a picture in the frame below. Use as many of the circles described on page 44 as possible.

Pi and Circumference

Pi is a constant that is used in many mathematical formulas and is represented with the Greek letter π. Pi is defined as the ratio of a circumference of a circle to the diameter of a circle ($\pi = C/d$). Complete the chart and do the following exercise to become familiar with the concept of pi.

Directions: Find and list on the chart below 10 circular items such as a clock, a wheel, etc. To find the circumference of each item, wrap it with string, and then measure the string with a ruler, yardstick, or meter stick. Measure the diameter of each item as well, and add this information to the chart. Finally, divide the circumference of each item by its diameter. The answer is the value of pi. (Note: Since you are not using precise measuring tools, your measurements may not be exact. Therefore, your answers may not be exactly the same. However, they should be close enough for you to approximate the value of pi.) Write your answer for the value of pi on the line below when you have finished the exercise.

Item	Circumference	Diameter	pi (or $\frac{C}{d}$)

Use the information above and the examples below to find the value of π. How close was your original calculation?

1. $C = 78.5$
   $r = 12.5$
   $\pi =$ ____________
2. $C = 196.25$
   $d = 62.5$
   $\pi =$ ____________
3. $C = 70.022$
   $d = 22.3$
   $\pi =$ ____________
4. $C = 39.564$
   $r = 6.3$
   $\pi =$ ____________

The value of pi is ______________________________.

Challenge: Here are two formulas for finding the circumference of a circle. Can you change them to show the formula for π?

$C = \pi d$, where C is the circumference and d is the diameter of the circle (The circumference is equal to the value of pi times the diameter.)

$C = 2 \pi r$, where C is the circumference and r is the radius of the circle (The circumference is equal to two times the value of pi times the radius.)

Area of a Sector

A sector is the part of a circle in which the boundaries are two radii and an intercepted arc. If you know the area of a circle (πr^2) and the number of degrees in the angle made by the sector ($n°$), can you find the area of that sector? Study the information in the example below.

Area of circle = 3600 cm^2

$n° = 60°$

Area of Sector = 600 cm^2

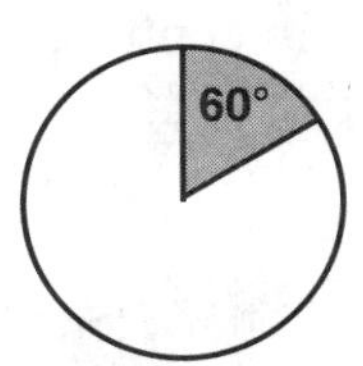

Formula for the area of a sector: $A = \frac{n°}{360°}(\pi r^2)$

Using the example above, if you know the sector angle ($n°$, or 60° in this example), you can then find the portion of the circle which is the sector ($60°/360° = 1/6$ of the total area of the circle). Once you find the total area of the circle (πr^2), you can then find $1/6$ of that total area to determine the area of the sector.

A (area of circle) = πr^2

$\pi = 3.14$

r = radius

A (area of sector) $= \frac{n°}{360°}(\pi r^2)$

$A = \frac{60°}{360°}(\pi r^2)$

$A = \frac{1}{6}$ of (3600 cm^2)

$A = 600\ cm^2$

If the area of the square is 3600 cm^2 as in the example above, the area of the sector is $1/6$ of that, or 600 cm^2.

Given the information below and the formula for the area of a sector, $A = \frac{n°}{360°}(\pi r^2)$, find the area of each sector.

1. sector angle = 18°
   radius = 3 cm
2. sector angle = 36°
   radius = 4 cm
3. sector angle = 90°
   radius = 5 cm
4. sector angle = 252°
   radius = 10 cm
5. sector angle = 270°
   radius = 10 cm
6. sector angle = 9°
   radius = 6 cm

Finding Areas Within Areas

Directions: Use the formulas for the area of a triangle ($A = \frac{1}{2}bh$) and the area of a circle ($A = \pi r^2$) to determine the total **shaded** area of each of the figures below.

Use the radii, measured in centimeters, to solve each problem. Make your computations, and write the answers in the boxes provided.

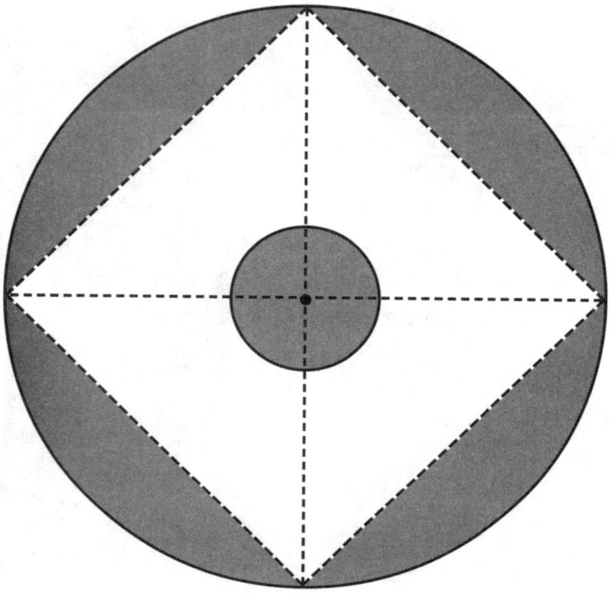

Computation:

Answer

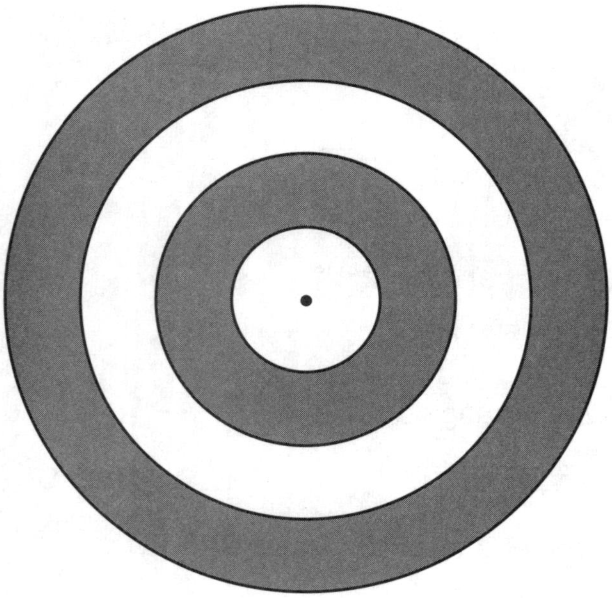

Computation:

Answer

Teacher Pages: Using Geometry

In this section, students will discover how to use geometry. Some exercises use more complicated forms of geometry, but most use basic concepts from which students can extend their thinking about the role of geometry in our lives.

The following information corresponds to the exercise pages in this section. Where applicable, teacher information that may be helpful to the students' understanding of the exercise is provided. For easy reference, the page numbers and titles of the exercises are given.

Page 51: Comparing in Geometry

The information about shape comparisons and the activities on this page help students to look more carefully at shapes. (Possible student solutions to the comparison activity at the bottom of the page include the following: cutting out the shapes and placing one over the other; measuring the sides or angles of shapes and comparing them; enlarging or reducing shapes to see if they are similar.) Students can then proceed with the activities on pages 52 through 59.

Pages 52 and 53: Congruence

Geometry books devote considerable space to congruence. In the activity on page 51, many students may have worked through the process of comparing shapes by cutting them out. While this method is concrete and definitive, it is not the easiest or most efficient way to make comparisons. The activities on page 52 use measurement as a means of comparison to prove congruence. When students have completed this page, have them create a mosaic design using congruent shapes (page 53).

Page 54: Similarity

Solutions:

1. (Figure ABCD) all sides = 1.5 cm, all angles = 90°; (Figure EFGH) all sides = 2.3 cm, all angles = 90°; similar (yes)
2. (Figure ABC) AB = 1.6 cm, BC = .5 cm, AC = 1.5 cm; (Figure DEF) DE = 4.8, EF = 1.5 cm, DF = 4.5; ABC and DEF angles are the same (B, E = 70°; C, F = 90°; A,D = 20°; similar (yes)
3. Figures ABCD and EFGH are not similar. Sides are not proportionate between figures. Angles are not the same.
4. Figures ABCD and EFGH are not similar. Angles are the same but sides are not proportionate between figures.

Page 55: Facts About Similar Triangles

Among the facts, or properties, students may discover are the following: the corresponding sides are in proportion; the corresponding angles are the same measure (congruent); the sides of an angle from one triangle are in proportion to the sides of the corresponding angle of the other triangle; in similar right triangles, an acute angle in one triangle is congruent to an acute angle in the other; the corresponding altitudes have the same ratios as the lengths of any pair of corresponding sides.

Page 56: Transformation

Solutions:

1. Figure ABC is rotated 90° clockwise.
2. Figure ABC is flipped along line segment BC, and then rotated 20° counterclockwise.
3. Figure ABC is rotated 54° counterclockwise.
4. Figure ABC is flipped along line segment AC.
5. Figure ABC is rotated 90° clockwise.
6. Figure ABC is rotated 90° counterclockwise.

Teacher Pages: Using Geometry (cont.)

Page 57: Identifying Symmetry

Lines of symmetry for figures 1–4 are as follows: 1. three lines, each passing through opposite vertices; 2. two lines, each dividing the rectangle in half; 3. The points of symmetry are the same as on last example of point symmetry on page 57.; 4. one line, passing vertically through center of heart.

Page 58: Symmetry By Reflection

Solutions: A line of reflection can be drawn between the pairs of figures in numbers 2, 3, 4, 5, and 6. The figures in numbers 1 and 4 represent the empty set.

Page 59: Symmetry By Translation

Solutions (movement of figures): **1.** 1 cm up; **2.** 1.3 cm up, .2 cm left; **3.** .6 cm up, 1 cm right; **4.** 1 cm up, .3 cm right

Page 60: Symmetry By Rotation

Solutions (Rotations): 1. 136°; 2. 180°; 3. 90°; 4. 55°

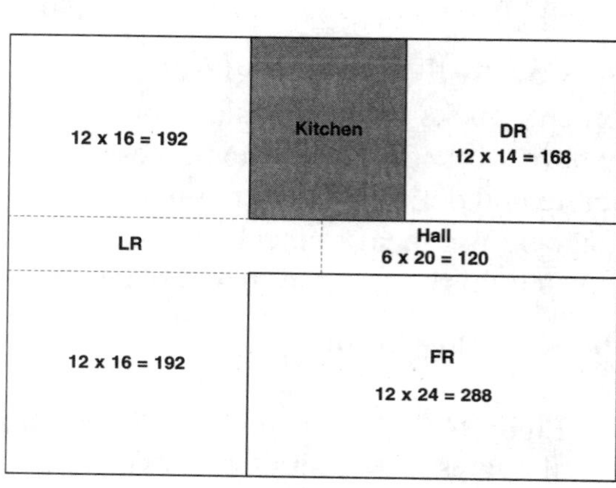

Page 63: Measuring Surface Area

Solution: 192 + 192 + 120 + 120 + 168 + 288 = 1200 ft.^{2}

Page 67: Creating With Tangrams

Tangram Animal Solutions:

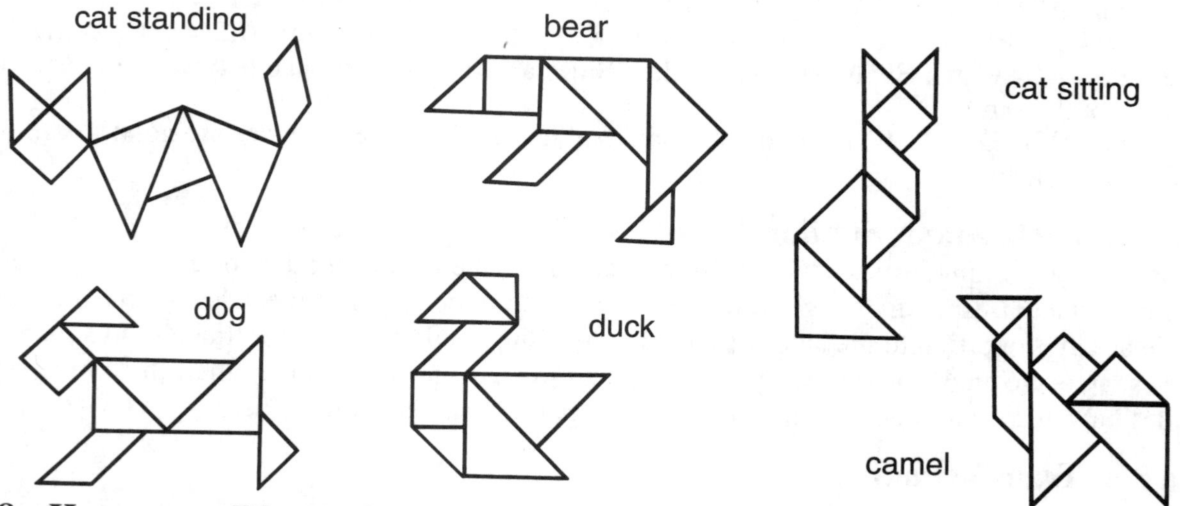

Page 68: How many Diagonals?

Solutions: 1. 5 sides, 5 vertices, 5 diagonals; 2. 6 sides, 6 vertices, 9 diagonals; 3. 7 sides, 7 vertices, 14 diagonals; 4. 8 sides, 8 vertices, 19 diagonals; challenge, 54 diagonals

Page 69: Square Versus Rectangle

The answer to this exercise is a square measuring 12 feet by 12 feet.

Comparing in Geometry

Points cannot be compared. They have no dimension. Lines, on the other hand, can be compared.

a ________ b ______________	a __________ b __________	b ______________ a ________
$a < b$	$a = b$	$b > a$

> This is called the *trichotomy principle*.

Shapes can be compared. The simplest shape is a triangle. We can often compare shapes such as the following just by observation.

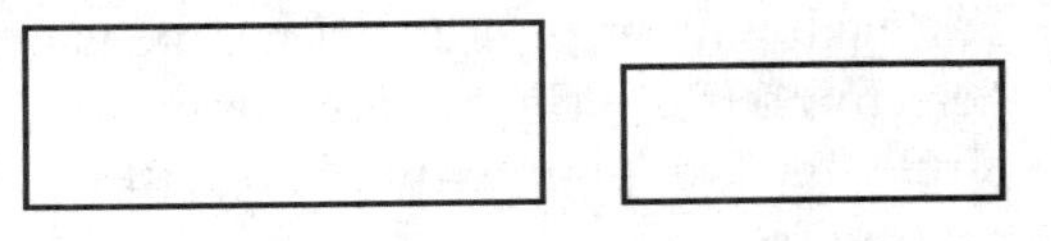

Shapes do not have to be side by side or in the same orientation to make comparisons. Shapes like the ones below are also easy to compare. Can you think of some possible ways to compare the following shapes to show which shapes are similar, which shapes are the same shape or size, or which shapes are dissimilar? Write your ideas on the back of this paper.

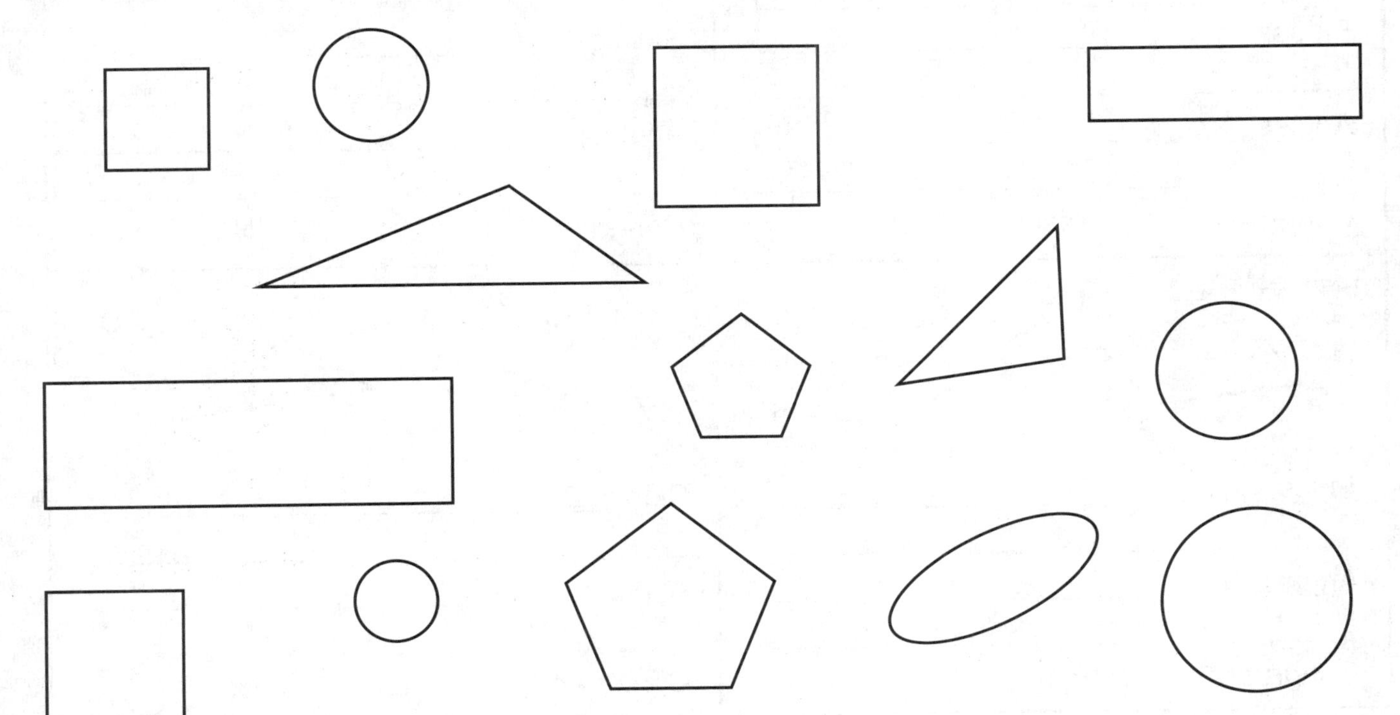

Congruence

In geometry shapes are compared in several ways. One method of comparison is to show that they are congruent. Figures that have the same size and shape are said to be congruent. When figures are congruent, all the sides and all the angles are correspondingly equal.

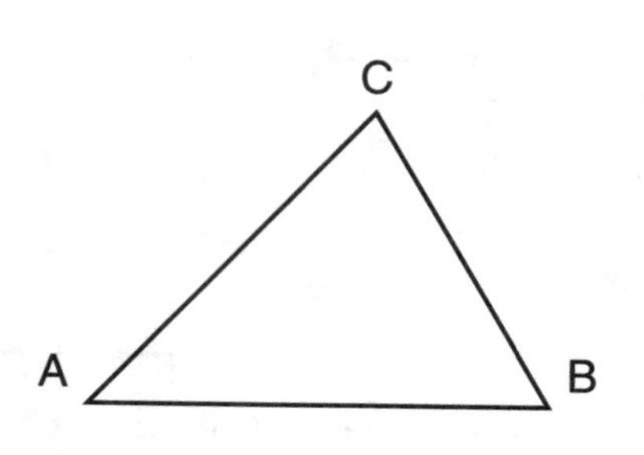

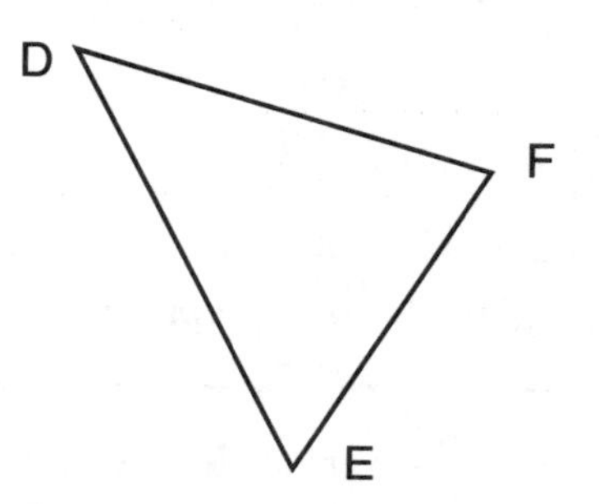

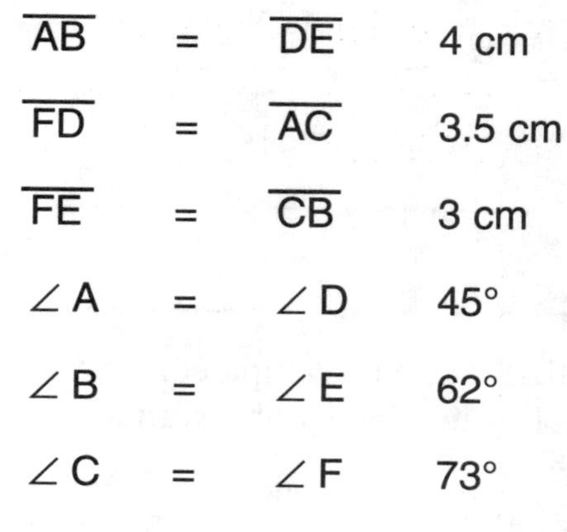

$\overline{AB}$	=	$\overline{DE}$	4 cm
$\overline{FD}$	=	$\overline{AC}$	3.5 cm
$\overline{FE}$	=	$\overline{CB}$	3 cm
$\angle A$	=	$\angle D$	45°
$\angle B$	=	$\angle E$	62°
$\angle C$	=	$\angle F$	73°
ΔABC	$\cong$		ΔDEF

(Congruent to)

Examples of congruent figures are all around you. Identical picture frames, books, windows, and bricks on a wall could be congruent.

For each pair of figures below, measure all of the sides and angles to determine if the figures are congruent. Circle "yes" or "no." If you circle "no", explain why. You will need a centimeter ruler and a protractor.

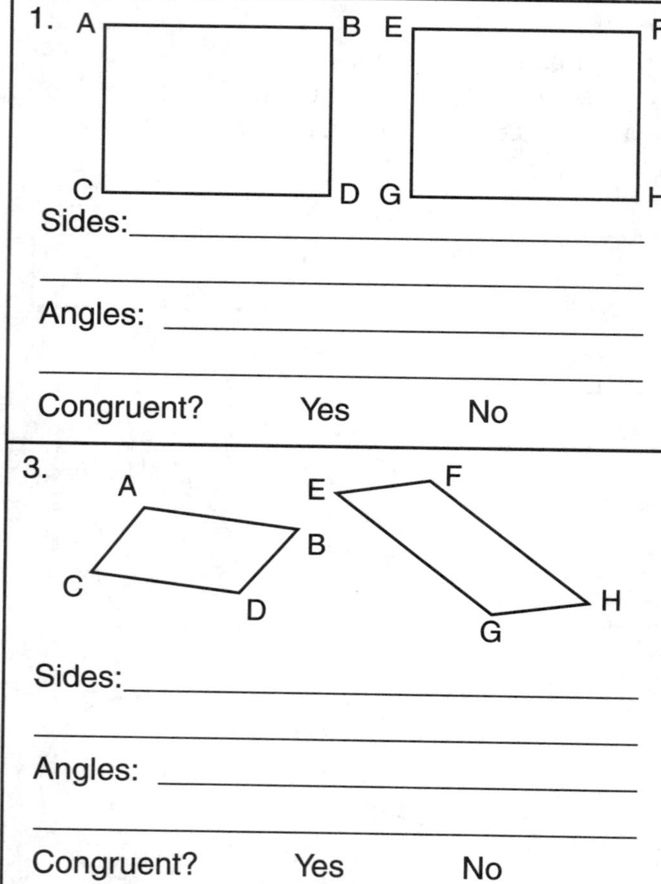

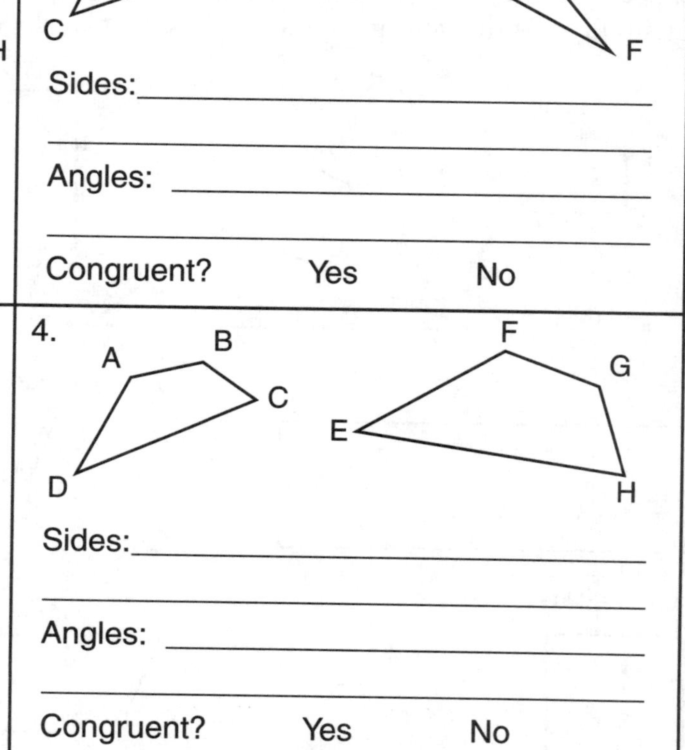

Congruence (cont.)

Directions: In the frame below, create a mosaic design that includes at least 20 pairs of congruent shapes. Fill the space with shapes and scatter the shapes that are congruent. The shape pairs do not have to be specific geometric shapes, such as regular polygons or circles, but they must be congruent. Share your completed design with someone. Challenge them to find all the congruent shapes.

Similarity

One way to compare geometric figures is to see if they are similar. Figures are similar if they have the same shape but not the same size. The sizes of the figures are proportionate.

> The symbol ~ is used to show that one figure "is similar to" another.

Figures such as triangles, squares, rectangles, and other polygons are similar if they have the same shape, corresponding equal angles, and sides that are correspondingly in proportion to each other.

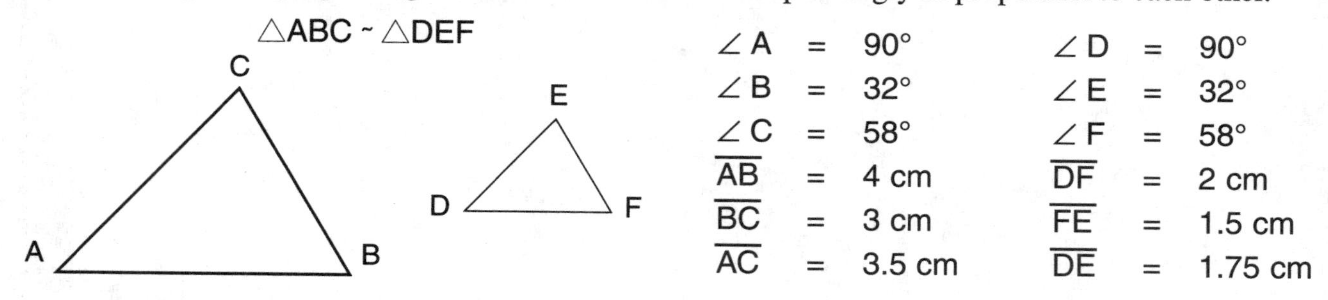

∠A	=	90°	∠D	=	90°
∠B	=	32°	∠E	=	32°
∠C	=	58°	∠F	=	58°
$\overline{AB}$	=	4 cm	$\overline{DF}$	=	2 cm
$\overline{BC}$	=	3 cm	$\overline{FE}$	=	1.5 cm
$\overline{AC}$	=	3.5 cm	$\overline{DE}$	=	1.75 cm

Directions: For each pair of figures below, measure all of the sides and angles to determine if the figures are similar. Circle "yes" or "no" to show whether the figures are similar. If you circle "no", explain why. You will need a centimeter ruler and a protractor.

1. A B C D E F H G

Sides: ______________________

____

Angles: ______________________

____

Similar? Yes No

2. A B C D E F

Sides: ______________________

____

Angles: ______________________

____

Similar? Yes No

3. A B C D E F G H

Sides: ______________________

____

Angles: ______________________

____

Similar? Yes No

4. A B C D E F G H

Sides: ______________________

____

Angles: ______________________

____

Similar? Yes No

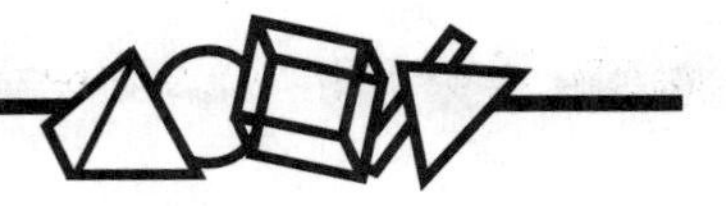

Facts About Similar Triangles

One fact about similar triangles is that they have the same shape but not the same size. However, there are other facts that you can discover by studying two similar triangles. Look carefully at the similar triangles in each of the boxes below. Use a centimeter ruler to measure the sides, and a protractor to measure the angles. Study the proportions, and so on, to see if you can discover other facts. Write your ideas on the lines provided. Use the back of this paper if you need more space.

Transformation

When a geometric shape is moved or rotated to a new position, we refer to this change as transformation. Study the figures below to see how the figure appears when it is transformed.

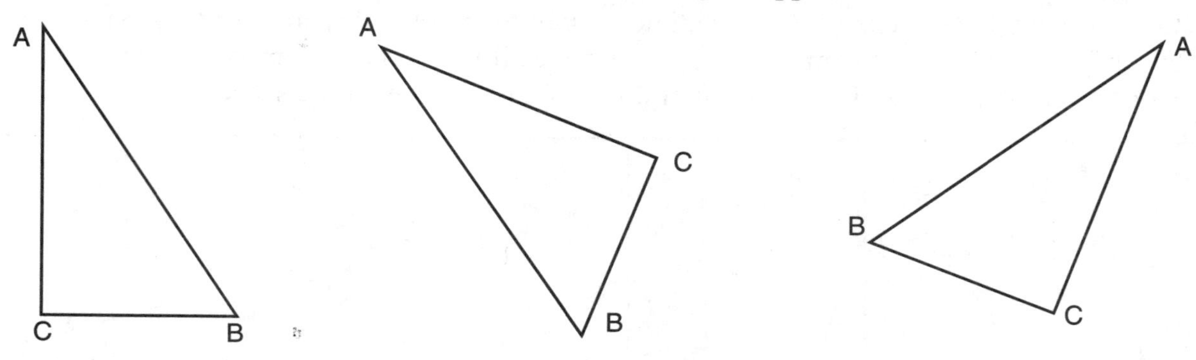

The triangle was flipped on side AB and shifted 90°. Look at the figures below. On the back of this paper write how each one was transformed. (Hint: Trace the original figure and cut it out. Then rotate the paper triangle in order to understand what is being done to transform it.)

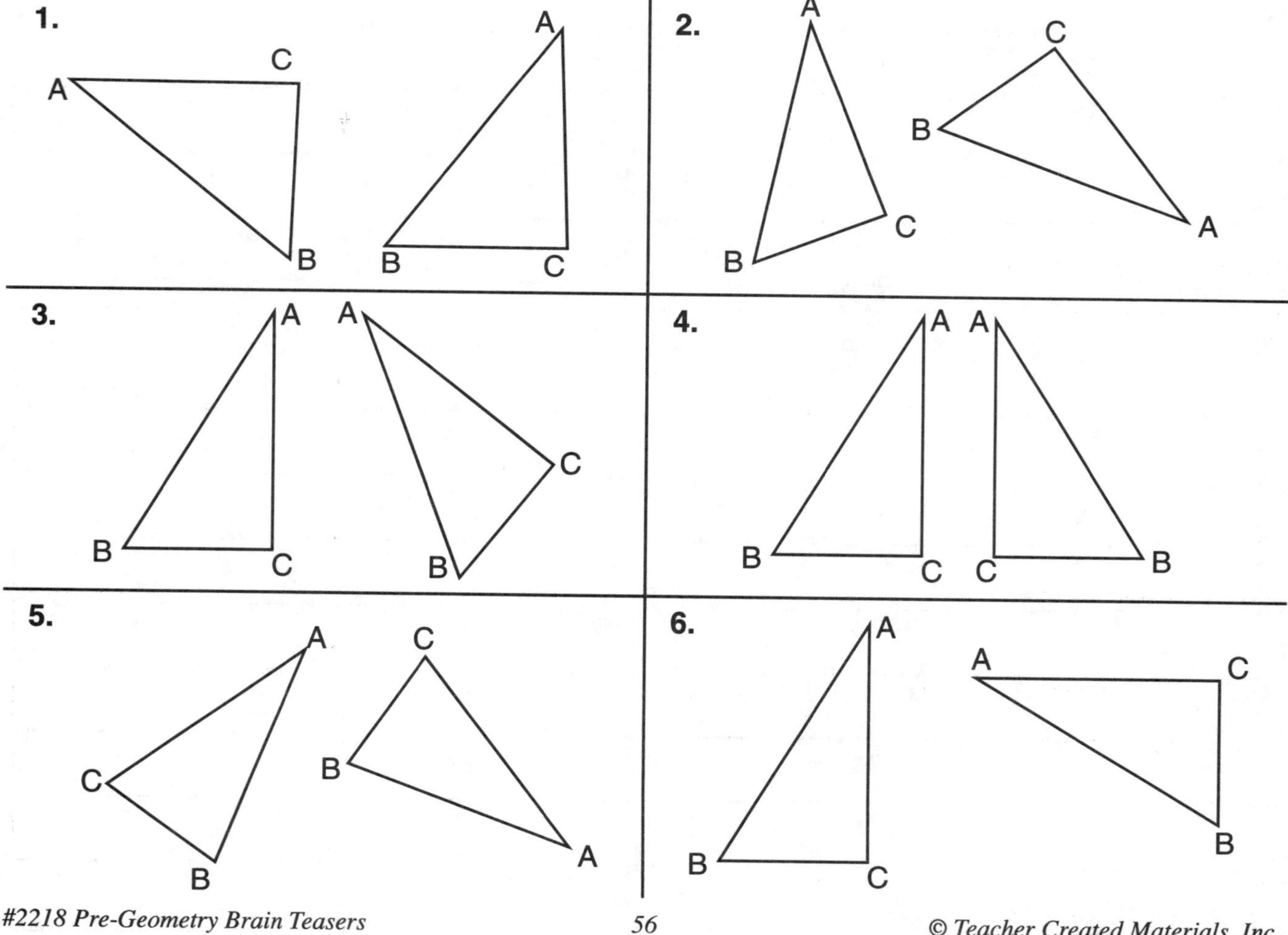

Identifying Symmetry

Symmetry occurs when a figure can be transformed onto itself. Squares and circles are always symmetrical. Point symmetry and line symmetry are two ways of identifying symmetrical figures.

When a line bisects a figure to create two congruent sides, the resulting symmetry is called reflection or bilateral symmetry. Here are some examples of reflection symmetry:

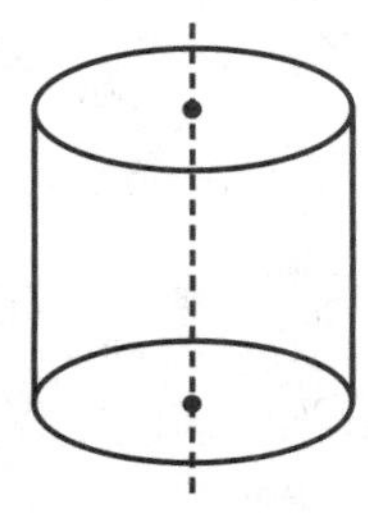

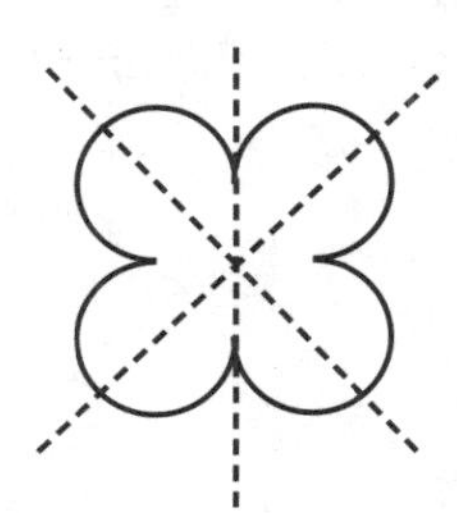

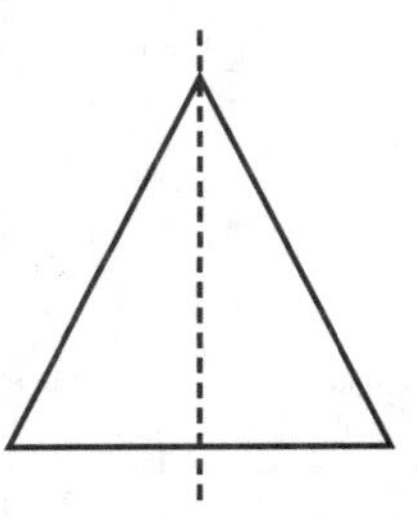

When everything in a figure radiates from a central point rather than a line, the resulting symmetry is called point symmetry. An easy way to understand point symmetry is to think of a cross section of an orange with a point at its center. Here are some examples of point symmetry:

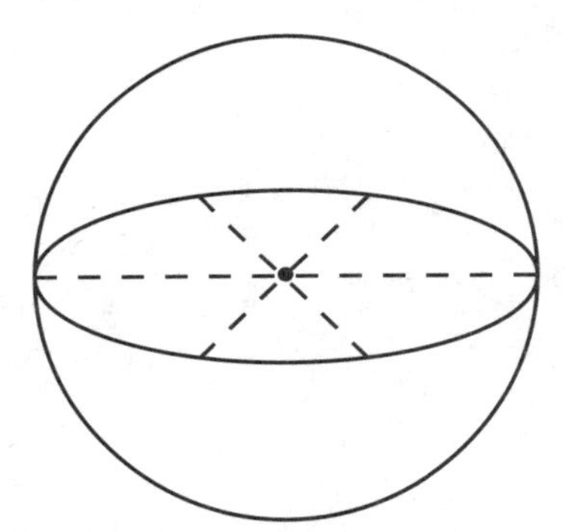

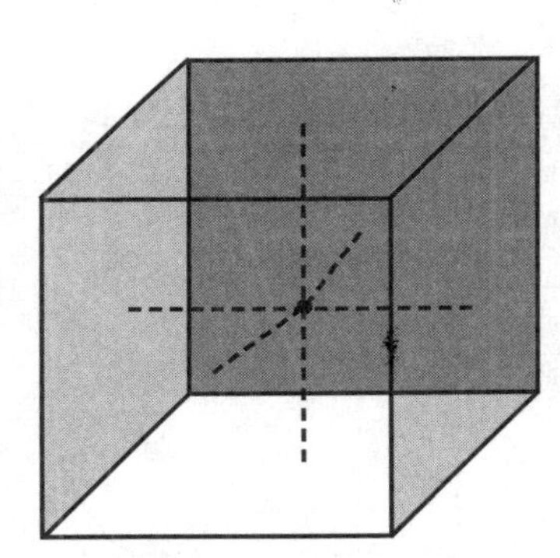

Directions: Draw lines to show the symmetry of the following figures and patterns:

1.	**2.**
3.	**4.**

Symmetry By Reflection

This special type of transformation occurs when all points are equidistant from a line. A line of reflection creates a mirror effect or mirror image. A reflection is also referred to as a flip. If an object is flipped, the resulting object is a mirror image, or reflection of the original object.

Directions: Study the figures below. If you are able to determine where the line of reflection is within a figure, use a straightedge to draw it. If not, write the empty set or null (0).

1.

2.

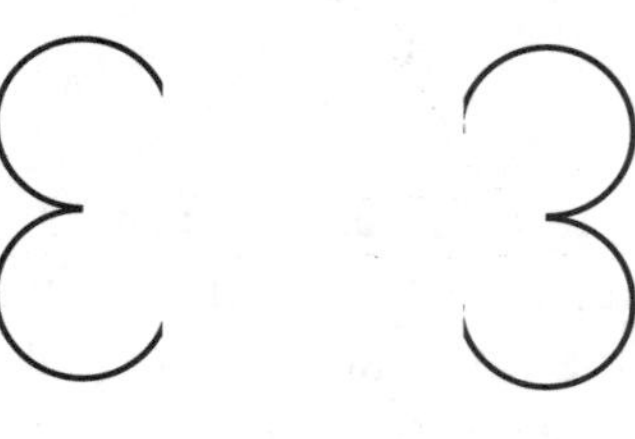

3.

4.

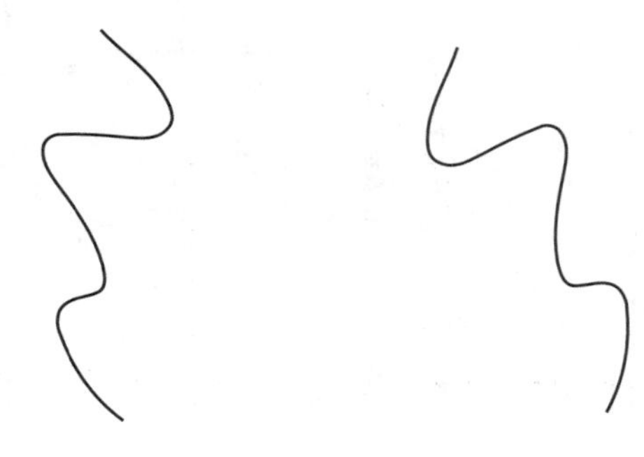

5.

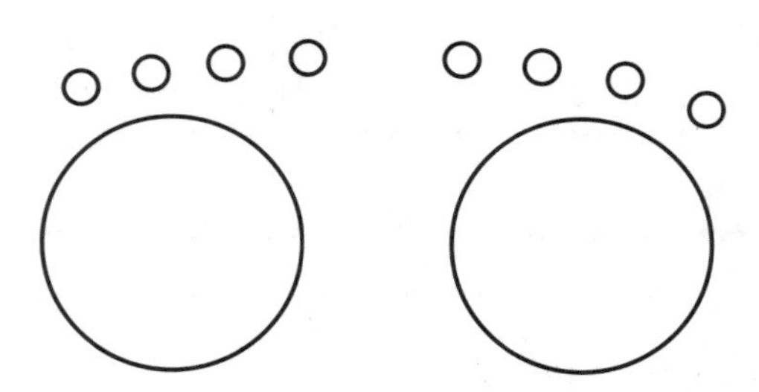

6.

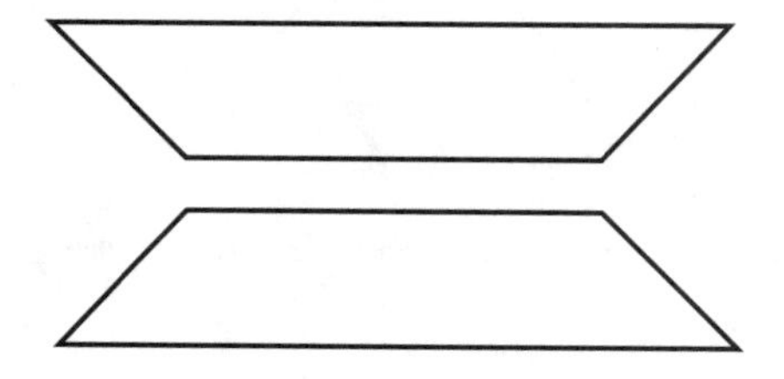

Challenge: Write the letters of the alphabet on a piece of paper. Draw the lines of symmetry where possible. Check each symmetry line by using a mirror, placed on the line, to see if the resulting shape is symmetrical. Which letters have one or more lines of symmetry?

Symmetry By Translation

A translation is a special type of transformation that occurs when a figure is translated or shifted a distance from its original position. A translation is also known as a glide since the figure glides to another position. You can describe the glide by noting the direction and distance of movement. In the translation below, the triangle has glided 5 centimeters to the right.

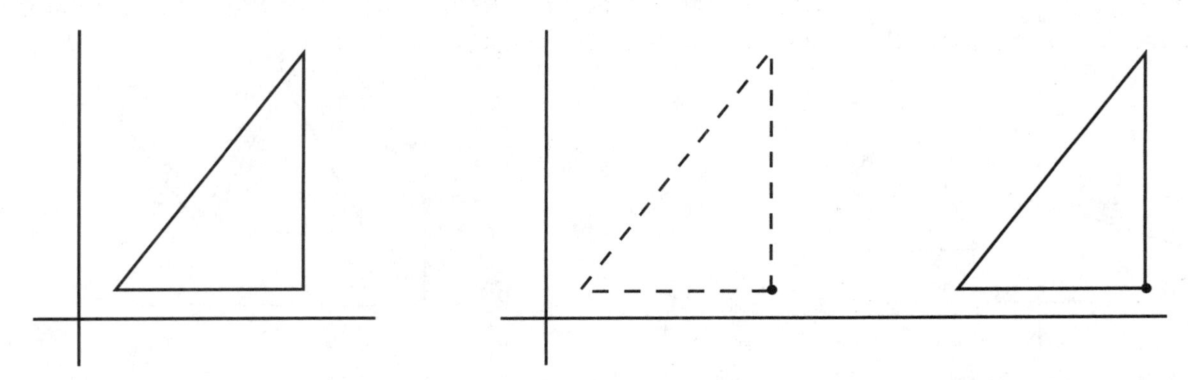

Directions: Describe the translation of each of the following pairs of figures. Use a centimeter ruler.

1.

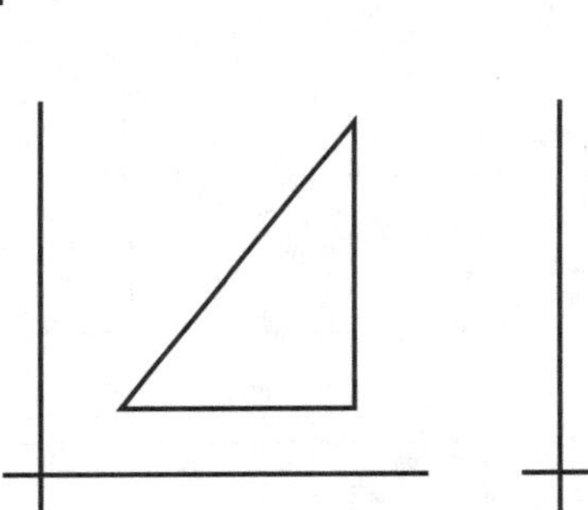

2.

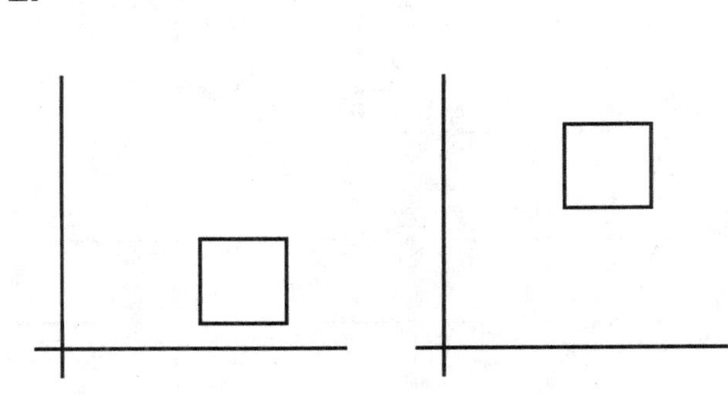

3.

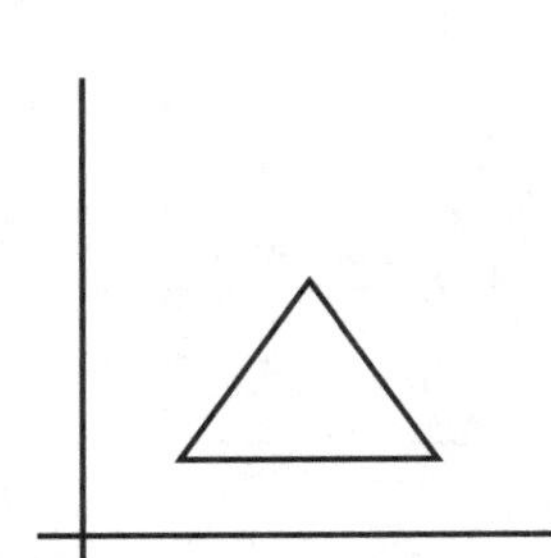

4.

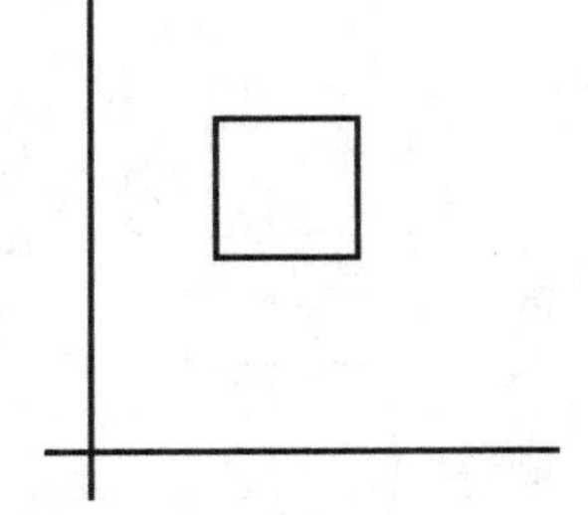

Challenge: On the back of this page, illustrate the following translations.

a. an isosceles triangle moved 4 centimeters to the right and 2 centimeters upward

b. a square moved 2 centimeters to the left and 4 centimeters downward

c. an equilateral triangle moved 3 centimeters to the right

Symmetry By Rotation

A rotation is a transformation that occurs when a figure is rotated about a point of symmetry. A rotation is also referred to as a turn since the figure turns through an angle of 360°. In the example below, triangle ABC rotates, or turns, 45° on point A. (Note: Side CA moves 45° on point A.)

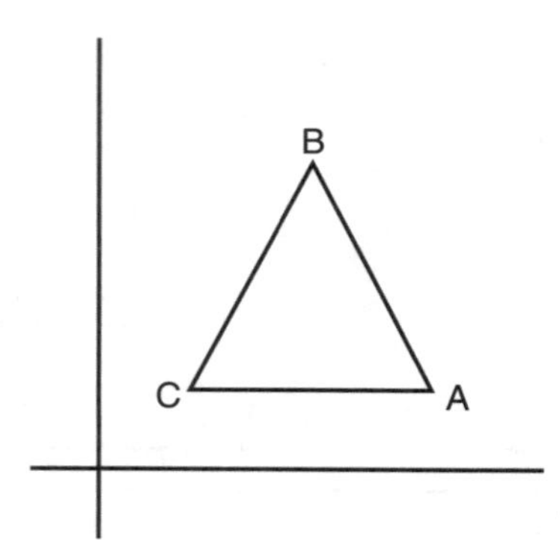

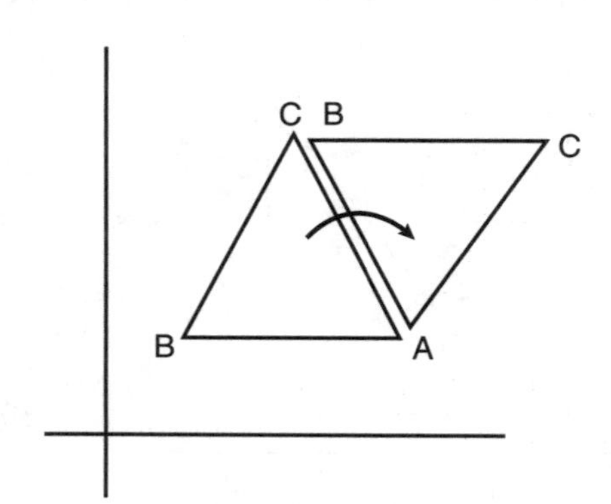

Directions: Measure the rotation for each of the figures below. Explain how the figure rotates.

1.

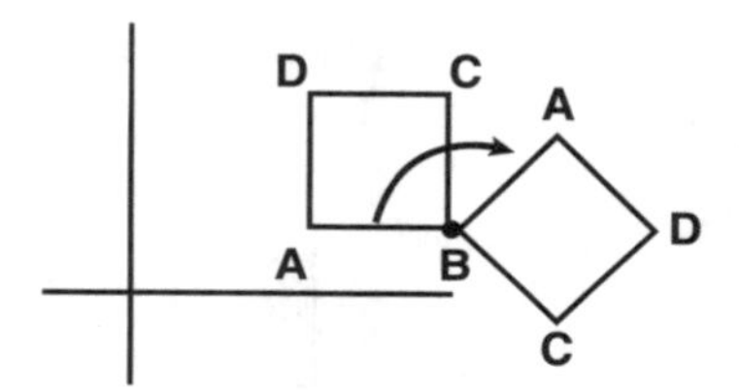

2.

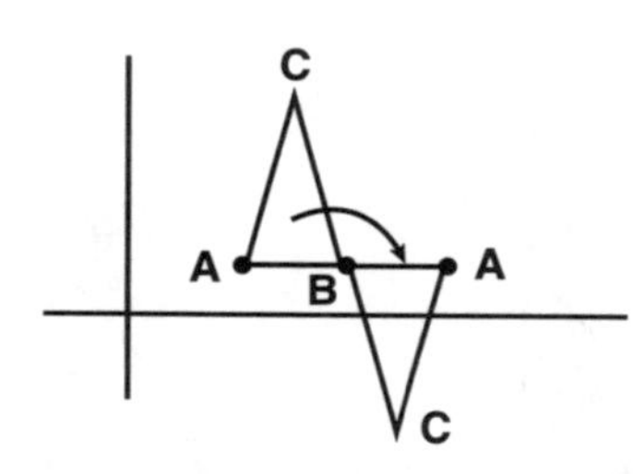

3.

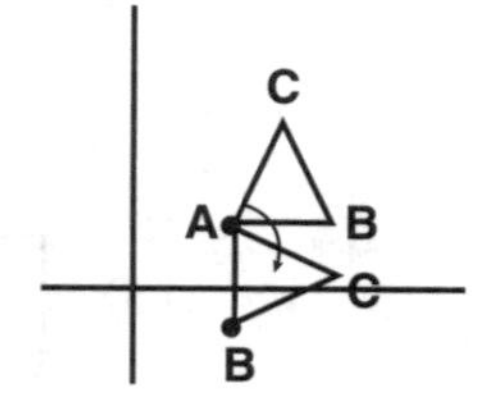

4.

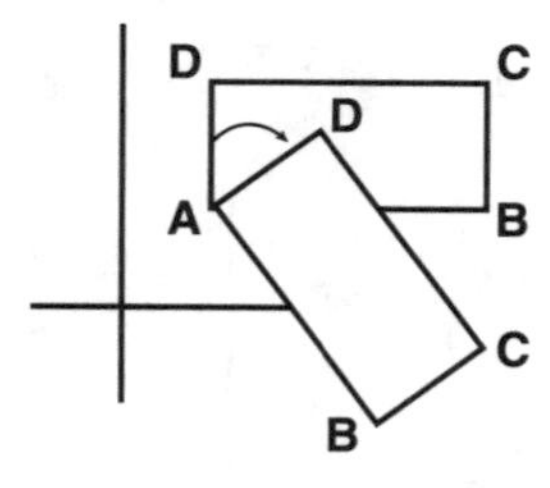

Challenge: On the back of this paper, illustrate the following rotations.

a. Rotate triangle ABC (scalene) 30° on point C.
b. Rotate square ABCD 90° on point D.
c. Rotate equilateral triangle ABC 100° on point A.

Tessellations

A tessellation is a design that uses interlocking shapes to fill a plane, as in a mosaic. Tessellations cover a plane and have no gaps or overlaps in the design. When a design is created in this way, the process is referred to as "tessellating" or "tiling" the plane. Tessellations can be made using various geometric shapes, block letters, or organic shapes.

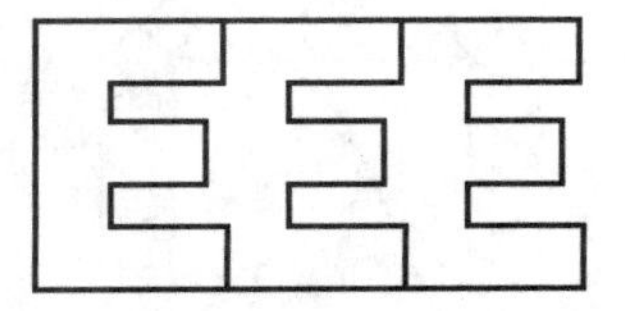

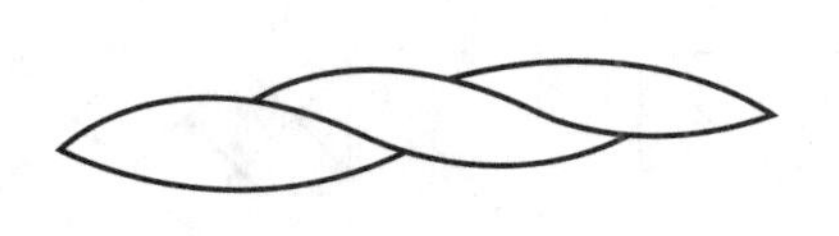

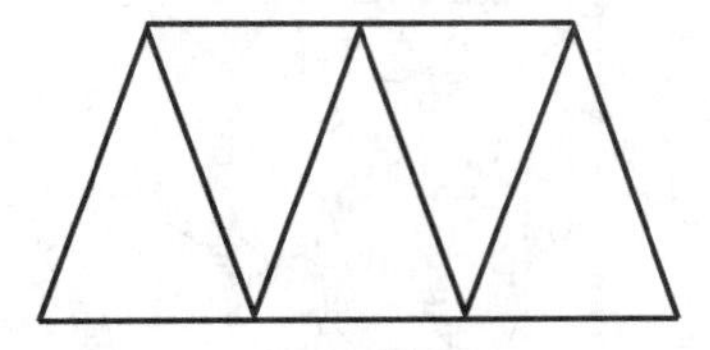

Directions: In each space below, create a tessellation. Use geometric figures in the first space, a block letter in the second space, and your own ideas in the other four blocks.

geometric figure	block letter

Designing Geometric Quilt Block Patterns

Directions: Use a compass, a ruler, and colored pens or pencils to create a different geometric design in each quilt block below. Make the design within each block symmetrical. Try to vary the block designs to make the most unique patterns possible.

Sample Block Designs

Measuring Surface Area

The measurement of geometric figures is not limited to the study of measurement in a geometry book. We measure geometric figures in every day situations, and we rely on our knowledge of surface area measurement to help us answer such questions as "How much paint (or wallpaper) do I need to cover the walls of this room?" or "How much fabric do we need to cover the backdrop for the stage?"

Use the diagram below to solve this problem for Jill. Jill needs new carpeting for the living room (LR), family room (FR), hall, and dining room (DR), which are on the first floor of her new house. The carpeting comes in 12-foot wide strips. As a result, Jill will have to purchase enough to allow for this. (A 12-foot wide strip of carpet will have to be seamed together with another 12-foot wide strip if one of the room's dimensions is more than 12 feet wide. She may have strips of carpeting left over that could be used elsewhere. All carpeting must run in the same direction)

Area of a Rectangle: A = l x w

Jill does not want to purchase more carpeting than she needs. What is the minimum number of square feet of carpeting Jill should purchase? Show your computations on the back of this paper.

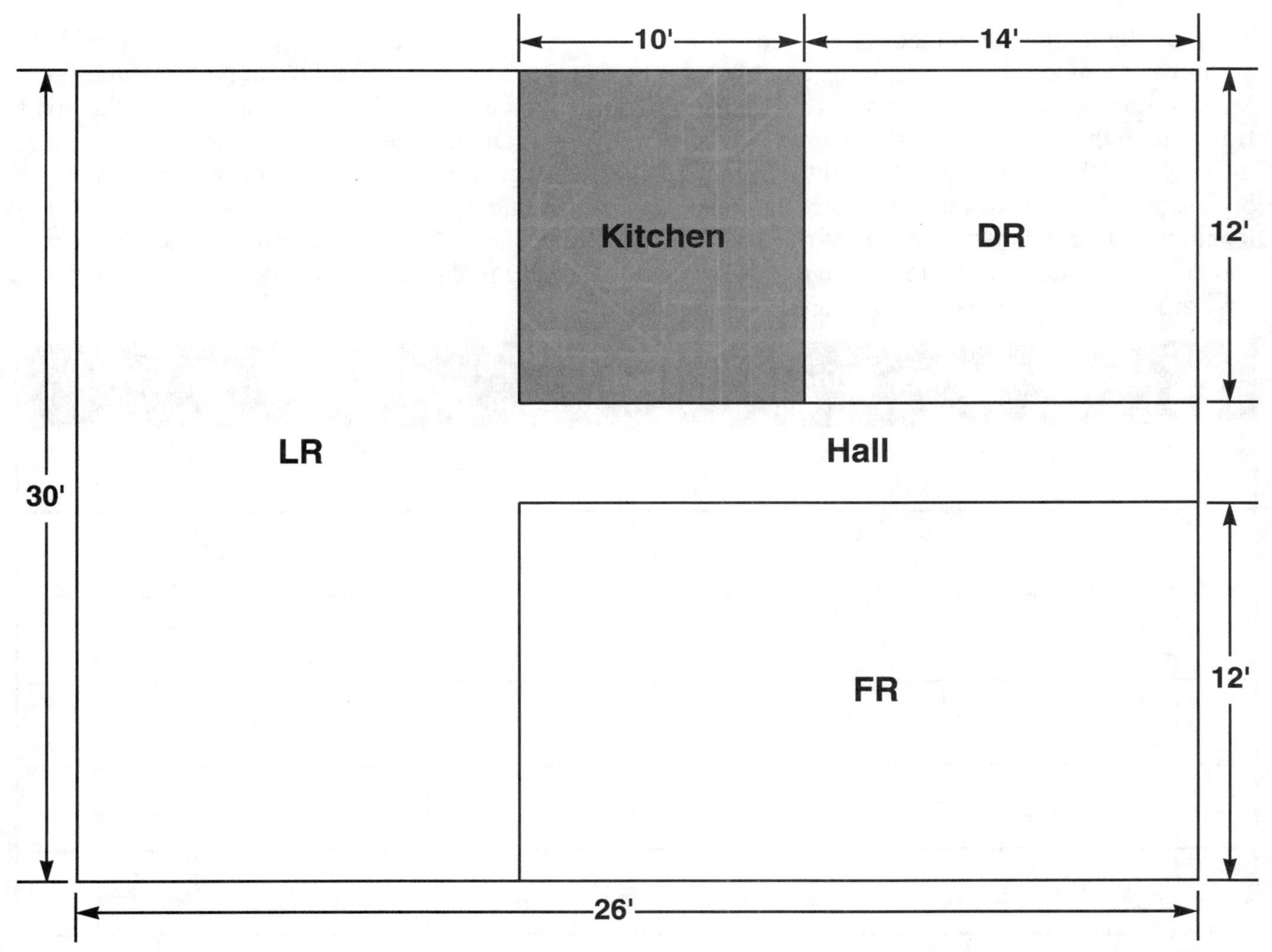

Measuring Distance

The Pythagorean theorem states that the sum of the squares of the legs of a right triangle is equal to the square of the hypotenuse. The formula is written as follows: $a^2 + b^2 = c^2$.

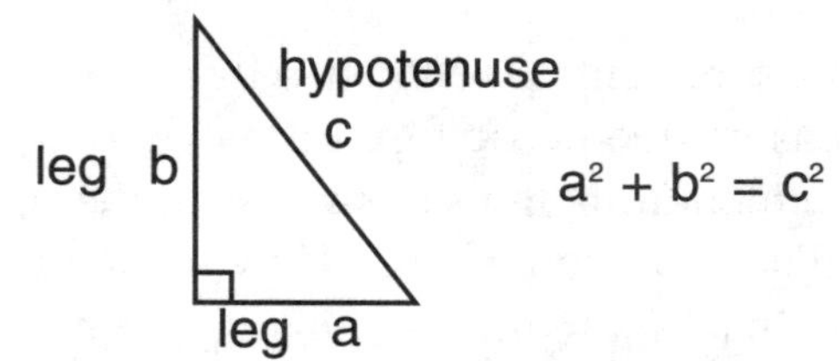

This formula can be used to measure distances. It is especially helpful when a measurement might otherwise be difficult to obtain.

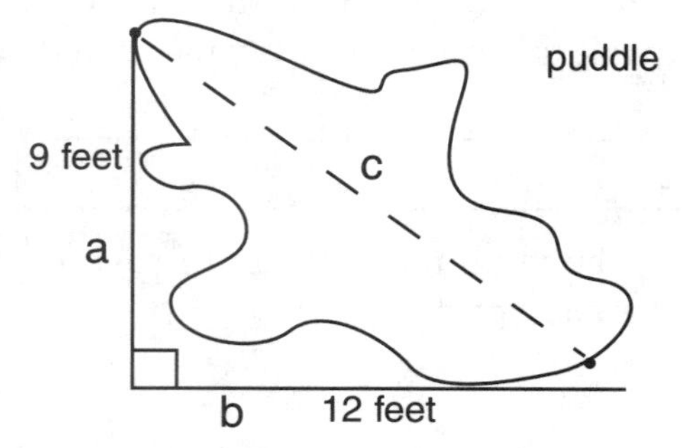

length of puddle = c

$$c^2 = 9^2 + 12^2$$

$$c^2 = 81 + 144$$

$$c^2 = 225$$

$$c = \sqrt{225}$$

$$c = 15 \text{ feet}$$

To apply the Pythagorean theorem, tie the endpoints of a string to a distance you want to measure. Write what you are measuring in the chart below. Pull the string out so that a 90° angle is created. Measure the leg of one of the rays of the right angle you have created. (That is, measure from the point from which the ray begins to the point at which it ends.) Subtract the measurement that you get from the length of the entire string to find the length of the other leg. You now have the measurements of both legs. These measurements represent a and b in the formula $a^2 + b^2 = c^2$. Use them to find the length of side c (the hypotenuse). Write your answers in the chart. Now measure five other distances and enter your information in the chart. Use your calculator to find each hypotenuse. (Round your answer for the hypotenuse to the nearest whole number.).

What I am measuring	length of leg a	length of leg b	length of leg c

Dream Catchers

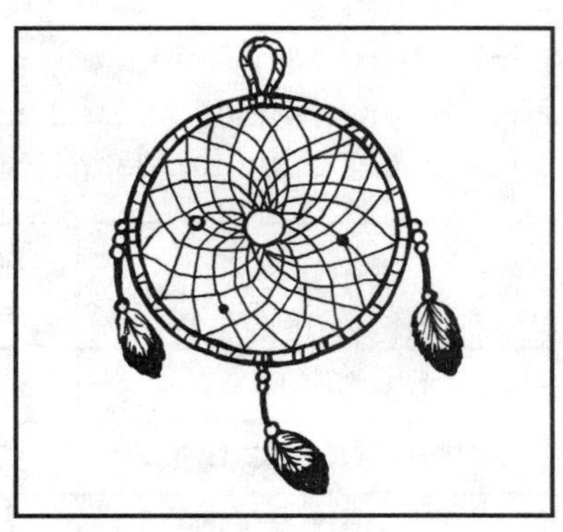

Dream catchers are special objects created by Native Americans to catch good dreams. Many dream catchers are circles woven with yarn or string to create a net inside. Special beads, shells, rocks, and feathers can be added to the net. Here is an example of a dream catcher.

Design a dream catcher pattern in the circle below. Use an interesting geometric pattern that could be used as a template for an actual dream catcher. Follow the pattern you designed to make your own dream catcher using a circular frame, string or yarn, and decorative materials to give it a unique look.

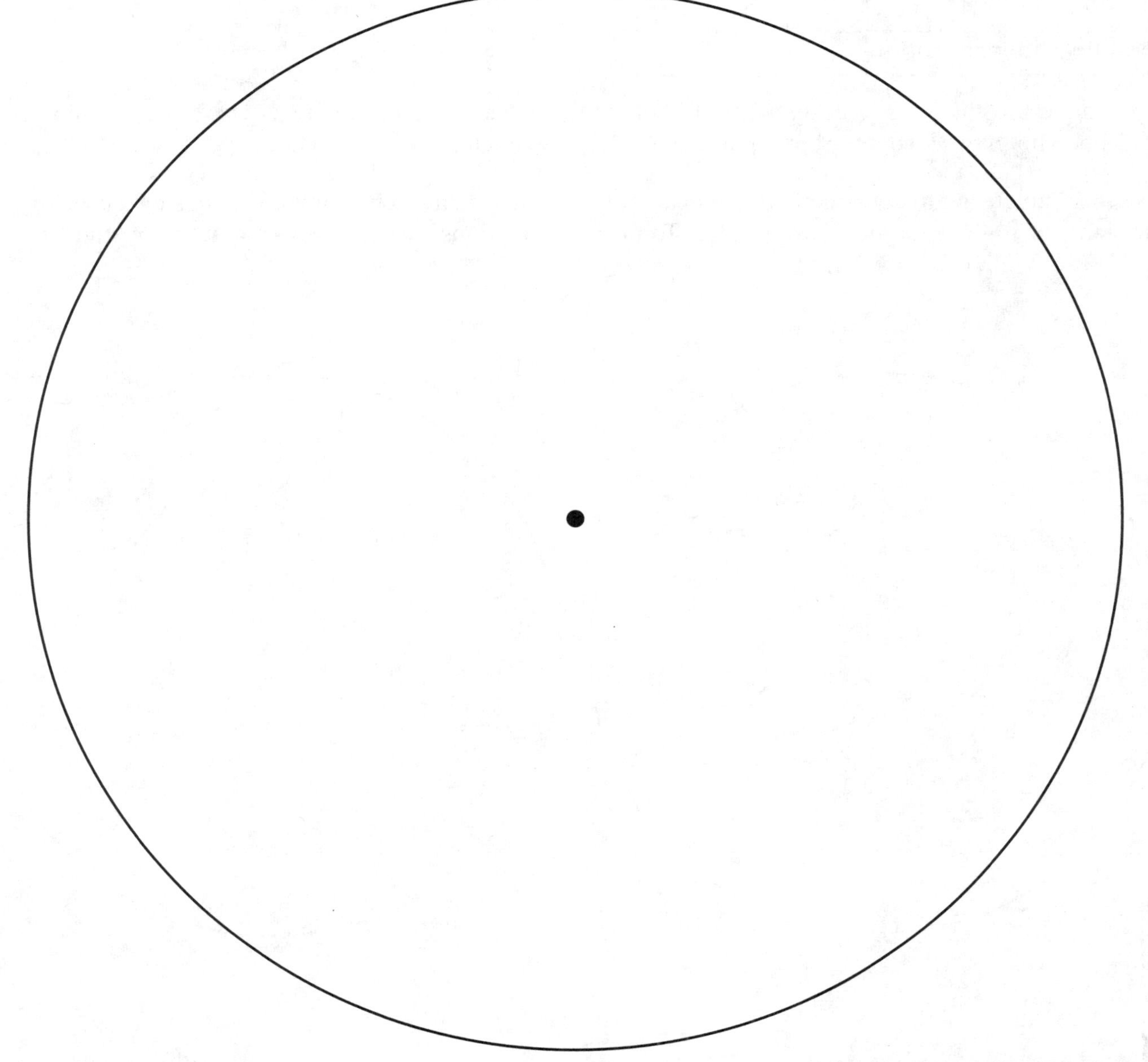

Pie Charts

A pie chart is a graphic way to display numerical information or data. Listed below is data about the activities performed by students in one class after school where n = 30 (students). A pie chart can be constructed using the data. (See illustration.)

Activity	Number of Participants
play soccer	10
attend dance class	5
do homework	4
go to the library	6
practice an instrument	5

Note that each student is represented by 12° of a circle because 360°/30 = 12°. Therefore, the ten students who play soccer are represented by 120° of the circle.

Use this knowledge to construct two pie charts of any other data, such as favorite foods or television programs of students in the class. (**Note:** To make calculations easier, choose a number of students that is evenly divisible by 360°.)

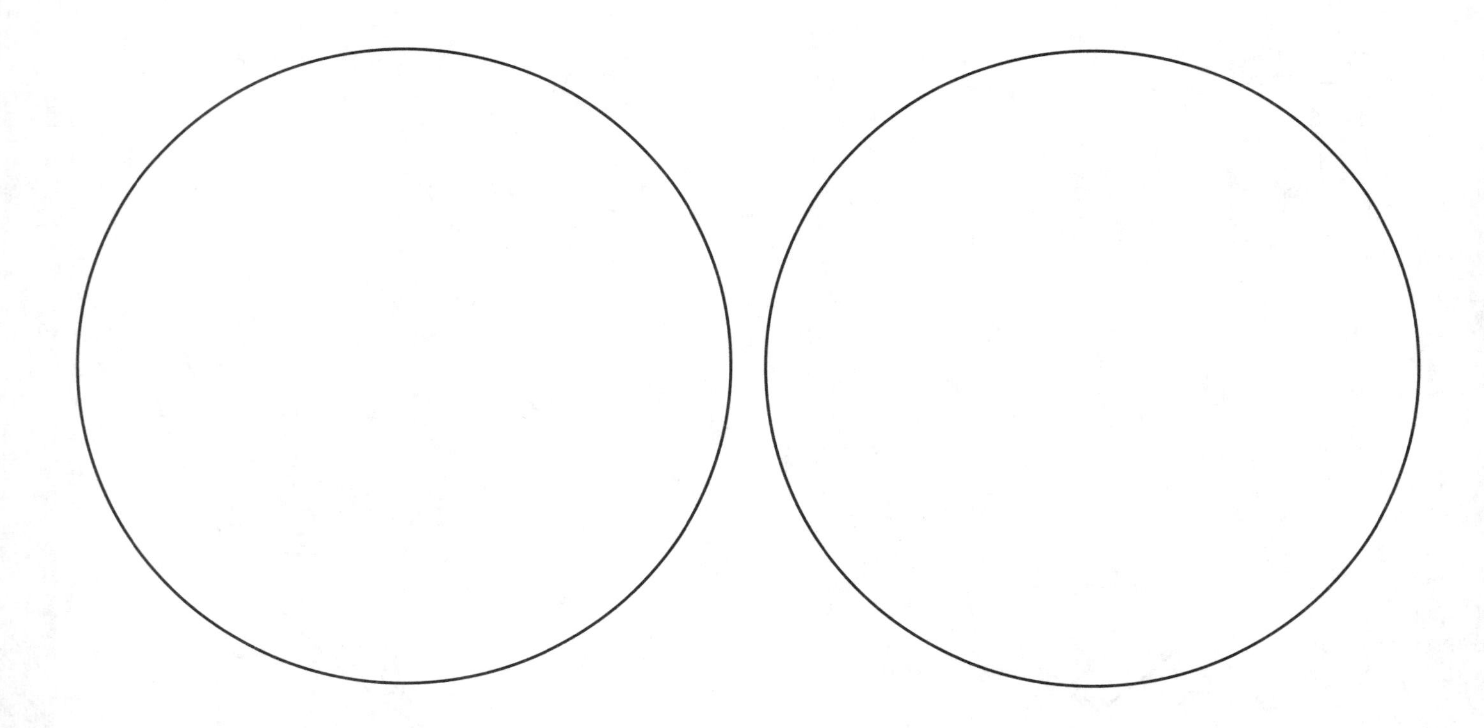

Tangrams

The tangram is a classic geometry puzzle found in many forms. Can you make the following animals using the tangram pieces? (You must use all pieces for each object or design.):

- cat (standing)
- cat (sitting)
- dog
- bear
- duck
- camel

Use the pieces to make as many objects or designs as possible.

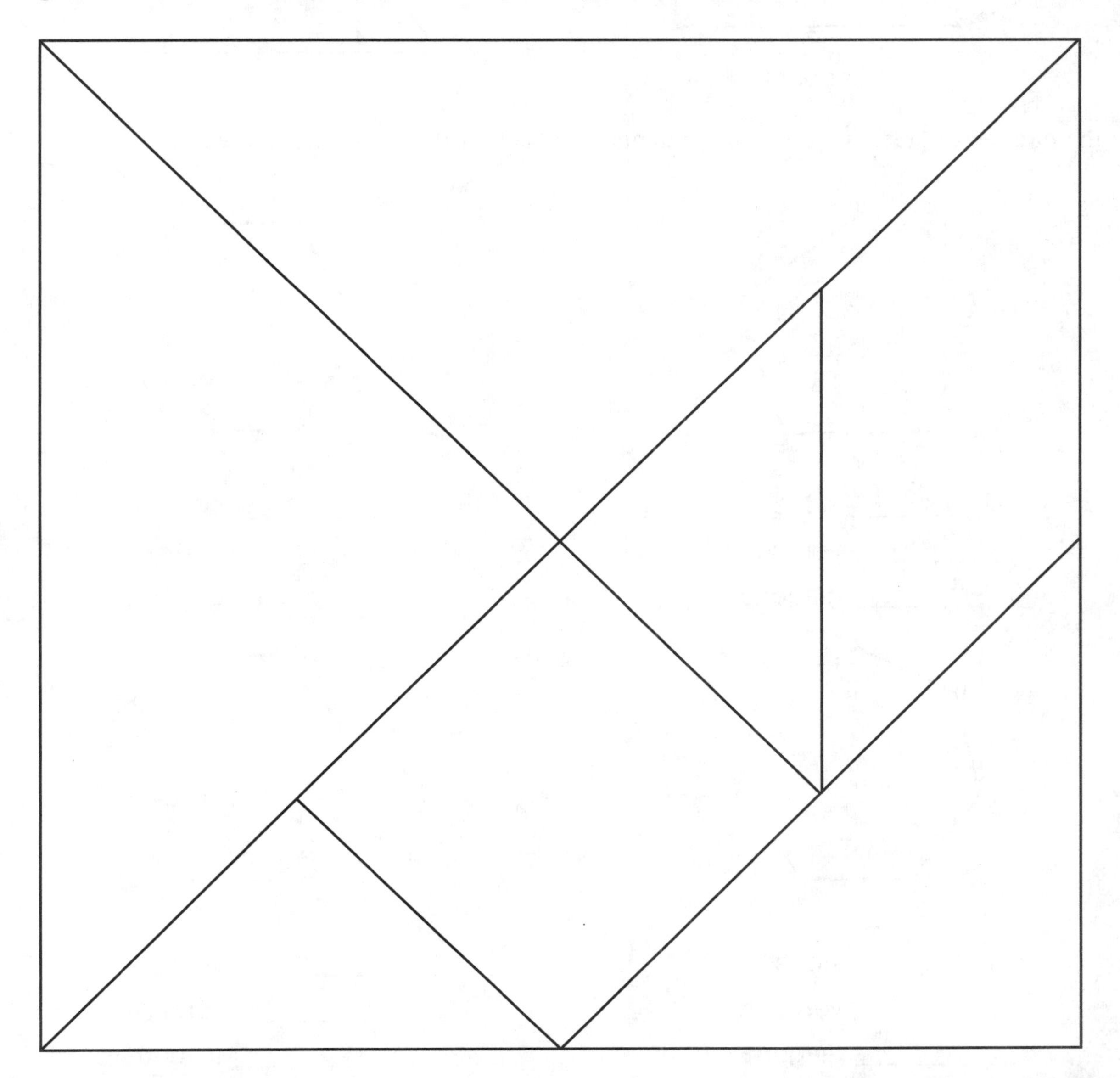

Challenge: Find the area of the tangram. Find the area of each of the pieces and add them to find the total. Does this total equal the area of the tangram? If your calculations are correct, the answers should be the same.

How Many Diagonals?

A diagonal is a line segment drawn between nonconsecutive vertices of a polygon. (Vertices are the points where two sides meet.) There are two diagonals in a rectangle and no diagonals in a triangle.

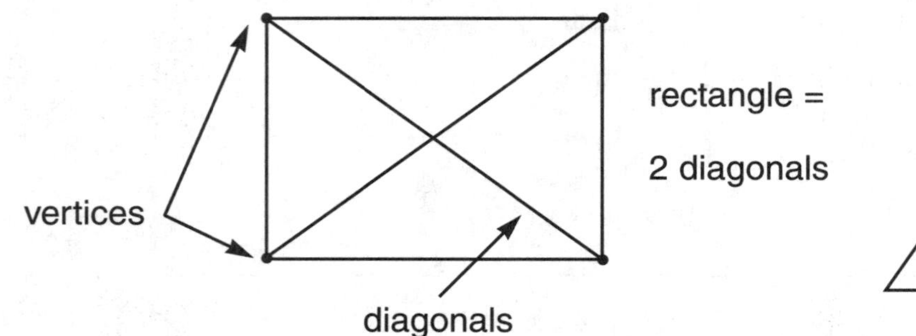

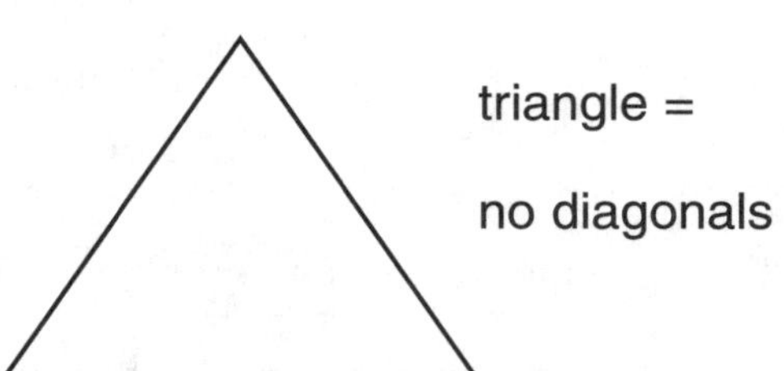

Write the number of sides, vertices, and diagonals in each of the following figures.

1.

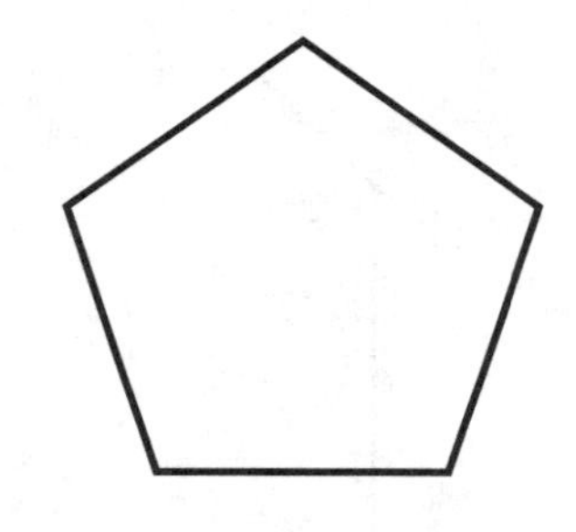

__________ sides
__________ vertices
__________ diagonals

2.

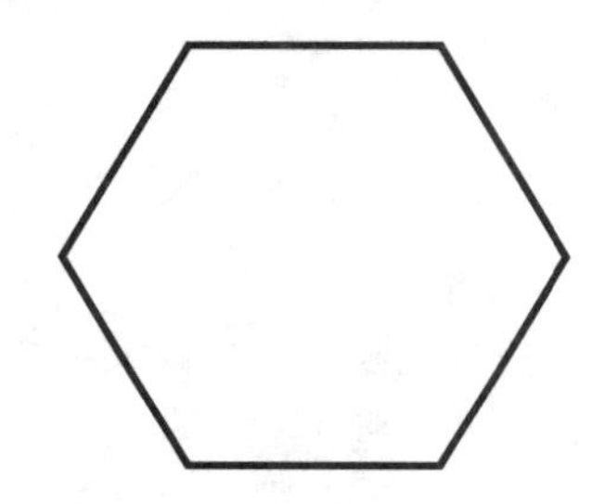

__________ sides
__________ vertices
__________ diagonals

3.

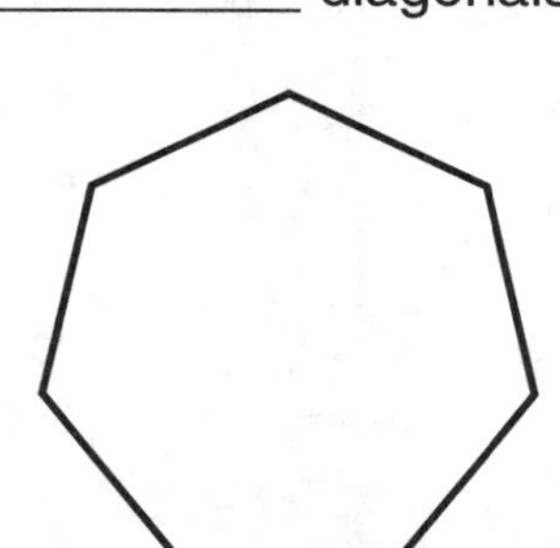

__________ sides
__________ vertices
__________ diagonals

4.

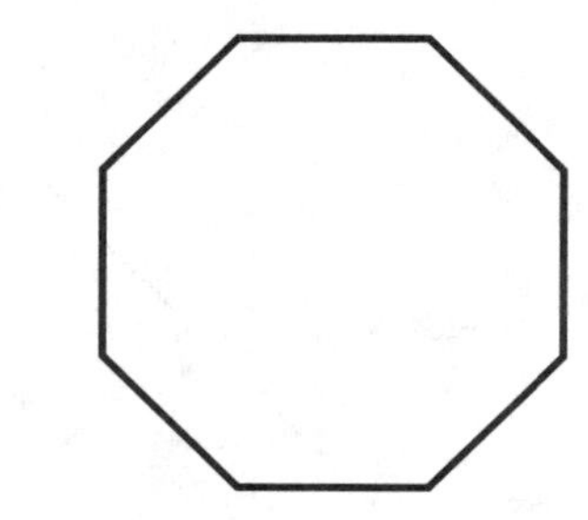

__________ sides
__________ vertices
__________ diagonals

Challenge: Draw a 12-sided figure. How many diagonals do you think it will have? Count all the diagonals possible. (Hint: Work with one vertex at a time.)

Square Versus Rectangle

In this activity, you are asked to compare the areas of two geometric shapes, a square and a rectangle.

At the local hardware store, a family purchased a 48-foot length of fencing to be used for a garden enclosure. The garden could be either a square or rectangular shape. The family has to decide on the most cost-efficient way to purchase the fencing. Which enclosure shape and dimensions will give them the greatest area? To answer this question, use the chart below. Include three scale drawings in which $^1/_8$ inch = 1 foot (.3 cm = 30 cm). Use the back of this page if you wish to try more scale drawings.

Scale Drawing	Length	Width	Area

Indirectly Measuring an Object

It is not always possible to directly measure an object. The object may be inaccessible or the tools to make a direct measurement may not be adequate. How can the object be measured? In the following problem, you will use trigonometry to determine the height of a tree.

Steps to Measuring the Height of a Tree

1. Find a tree and measure the distance from the point at which you will be standing to its trunk.

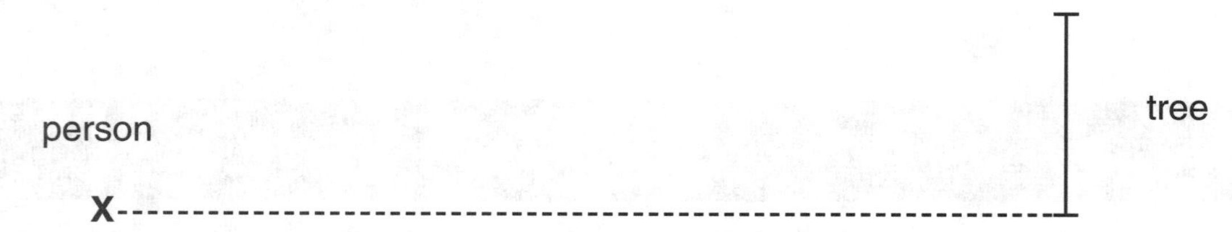

2. Stand at that point with a protractor in hand. Hold the protractor parallel to the ground and look up to the top of the tree. Note the angle of inclination.

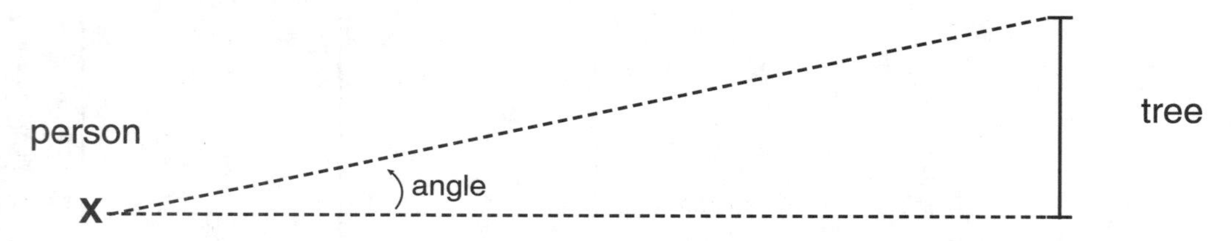

3. You now have an angle and a distance. Take a piece of paper and make a triangle that takes into account the positions of you and the tree. Be as accurate as possible.

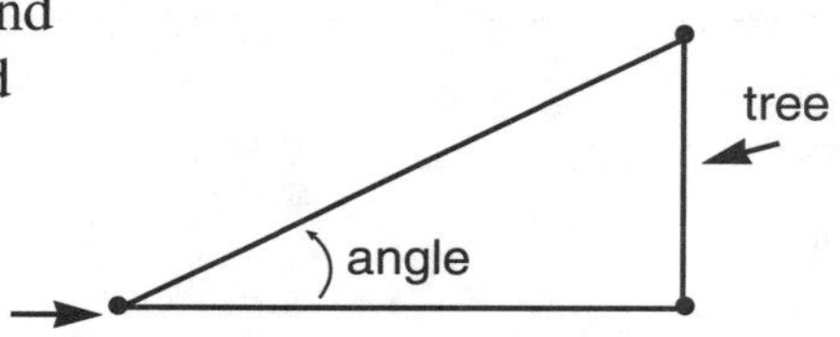

4. Set up the following proportion:

$$\frac{\text{paper tree}}{\text{distance on paper}} = \frac{\text{actual tree}}{\text{distance from tree}}$$

$$\frac{2 \text{ cm}}{4 \text{ cm}} = \frac{n}{40 \text{ m}}$$

5. You will notice that the height of the actual tree is equal to the height of the paper tree times the distance from the actual tree, divided by the distance on paper.

$$n \text{ (height of tree)} = \frac{2 \text{ cm} \times 40 \text{ m}}{4 \text{ cm}} = 20 \text{ m}$$

Measure more trees or some other tall objects. Make a chart on the back of this paper to show the measurements and how you solved the problems.

Teacher Page: Solid Geometry

The activities in this section primarily involve the examination of solid figures and the determination of volume as it relates to solids.

The following information corresponds to the exercise pages in this section. Where applicable, teacher information that may be helpful to the students' understanding of the exercises is provided. For easy reference, the exercise page numbers and titles of the exercises are given.

Page 72: Polyhedra

In this activity, students study the group of solids known as polyhedra. They are asked to identify the face, edge, and vertex and to guess the number of each in several polyhedra. At some point, explain the derivation of the terms tetrahedron, pentahedron, hexahedron, etc. This will help students remember the number of faces for each polyhedron.

tetrahedron —4
pentahedron—5
hexahedron—6
heptahedron—7
octahedron—8
nonahedron—9
decahedron—10
undecahedron—11
dodecahedron—12
icosahedron—20

Pages 73 and 74: A Closer Look at Polyhedra

The patterns provided on pages 75–78 can be used to help students count the number of faces, edges, and vertices for the polyhedra discussed on page 73. Students will need to cut out and assemble the polyhedra on pages 75–78. They should fold patterns along the dashed lines and then fold and glue the tabs to form the figure. Once they have constructed the tetrahedron, hexahedron, octahedron, dodecahedron, and icosahedron, have students complete the chart. The formula for finding the number of vertices, faces, and edges in a polyhedron is provided on the page. Euler's Formula can be used to complete the chart on page 74.

Use this chart for the solutions to pages 73 and 74.

Polyhedron	Faces	Vertices	Edges
tetrahedron	4	4	6
pentahedron	5	5	8
hexahedron	6	8	12
heptahedron	7	10	15
octahedron	8	6	12
decahedron	10	12	20
dodecahedron	12	20	30
icosahedron	20	12	30

Page 80: Finding Volume

Solutions: 1. volume of Peterson's pool = 4,521.6 cubic feet, volume of Andrew's pool = 4,500 cubic feet (Peterson's pool holds the most water.); 2. volume of each cylinder = 125.6 cubic centimeters (125.6 x 5 = 628), volume of rectangular prism = 1250 cubic centimeters, volume of cube = 216 cubic centimeters (total volume = 2094 cubic centimeters)

Polyhedra

Many geometric figures have only two dimensions, length and width. Such figures are studied in plane geometry, which deals with figures or surfaces having two dimensions on a plane. Solids, on the other hand, add a third dimension, depth, and are studies in solid geometry. One important group of solids is called *polyhedra* (the plural of polyhedron). Polyhedra are formed by the flat surfaces of polygons. Each of the flat surfaces is called a *face*. The place where two faces intersect is called an *edge,* and the point where three or more edges intersect is called a *vertex*.

Polyhedra are classified by the number of faces they have.

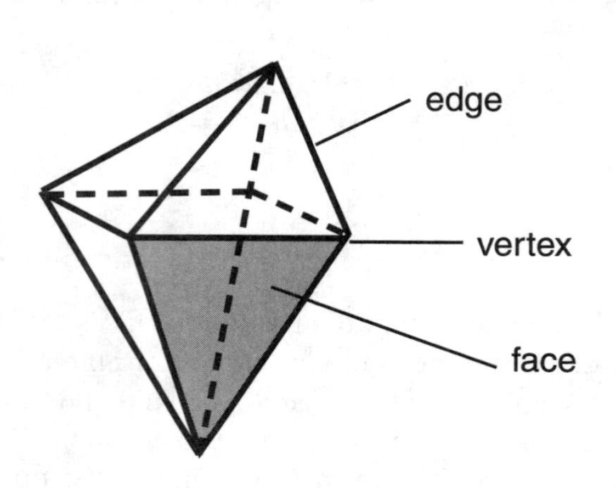

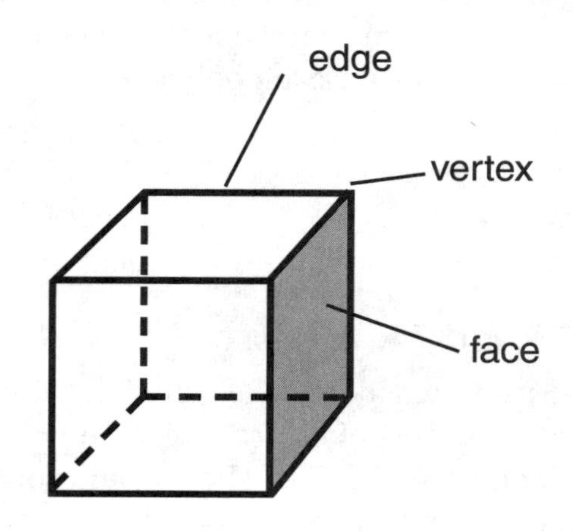

Common Polyhedra

- tetrahedron
- pentahedron
- hexahedron
- heptahedron
- octahedron
- nonahedron
- decahedron
- undecahedron
- dodecahedron
- icosahedron

Look at the chart on page 73. Can you guess how many faces, vertices, and edges each polyhedron has? Solve this problem by cutting out the patterns on pages 75–78 to construct each polyhedron. Then, complete the chart with the information you discovered as a result of your constructions.

A Closer Look at Polyhedra

Directions: After you have constructed each polyhedron from pages 74–78, complete the chart below.

Polyhedron	Faces	Verticles	Edges
tetrahedron			
hexahedron			
octahedron			
dodecahedron			
icosahedron			

Look at your completed chart. Can you find a relationship between the number of vertices, edges, and faces for each polyhedron? A Swiss mathematician named Leonard Euler discovered a relationship and created a formula for polyhedra. It is known as Euler's formula.

Euler's Formula $V + F - 2 = E$

V represents the number of vertices.
E represents the number of edges.
F represents the number of faces.
Use this formula to check the information in your chart.

A Closer Look at Polyhedra (cont.)

The following chart contains other polyhedra. Can you use Euler's formula to determine the number of faces, vertices, and edges for each figure? Fill in the chart.

Euler's Formula: V + F – 2 = E

Polyhedron	Faces	Vertices	Edges
pentahedron			
decahedron			
heptahedron			

Polyhedron Patterns

Directions: Cut out the shapes. Fold down all the dashed lines. Connect the sides of the polyhedron by matching and gluing each tab to its correct edge.

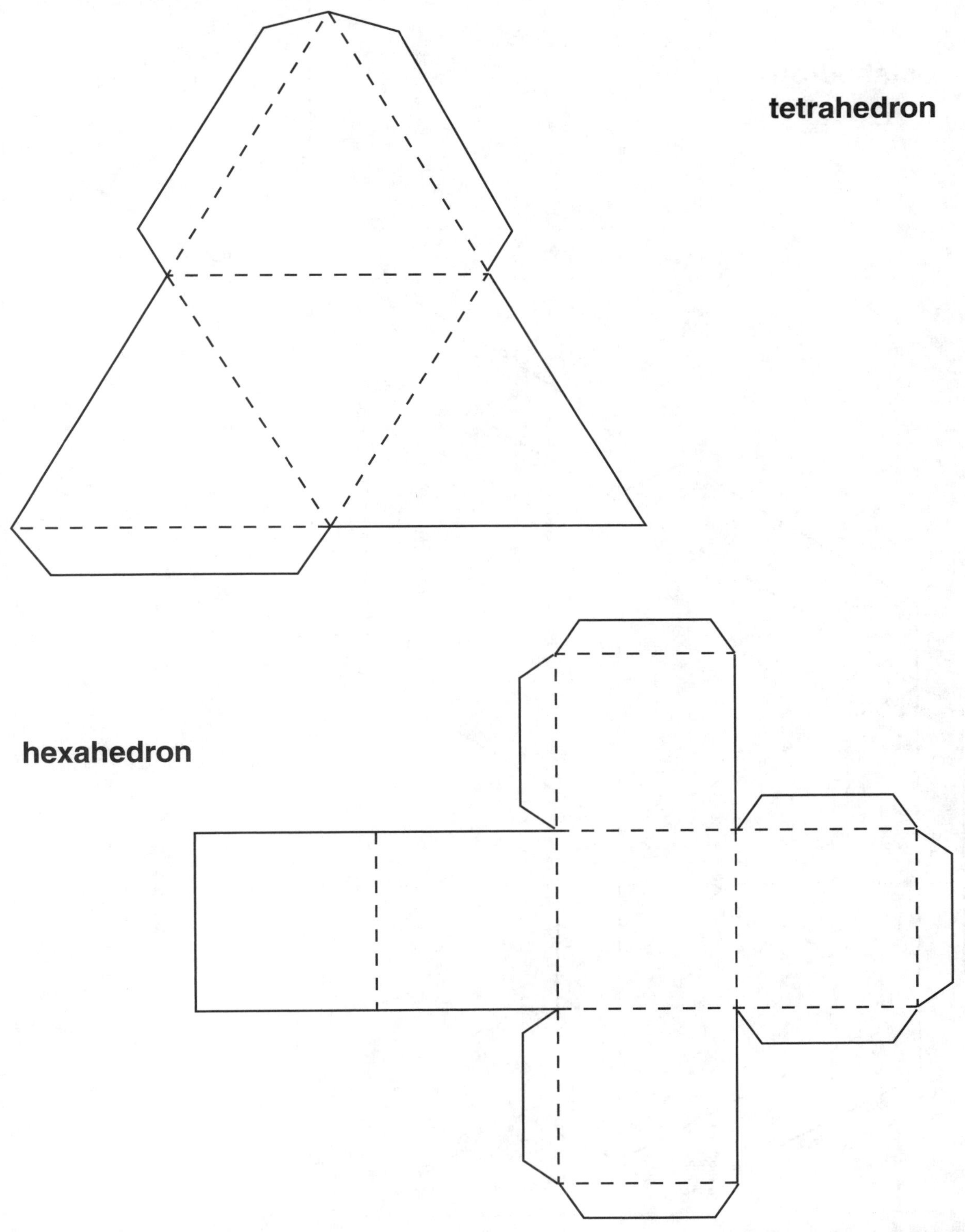

Polyhedron Patterns (cont.)

See page 75 for directions.

octahedron

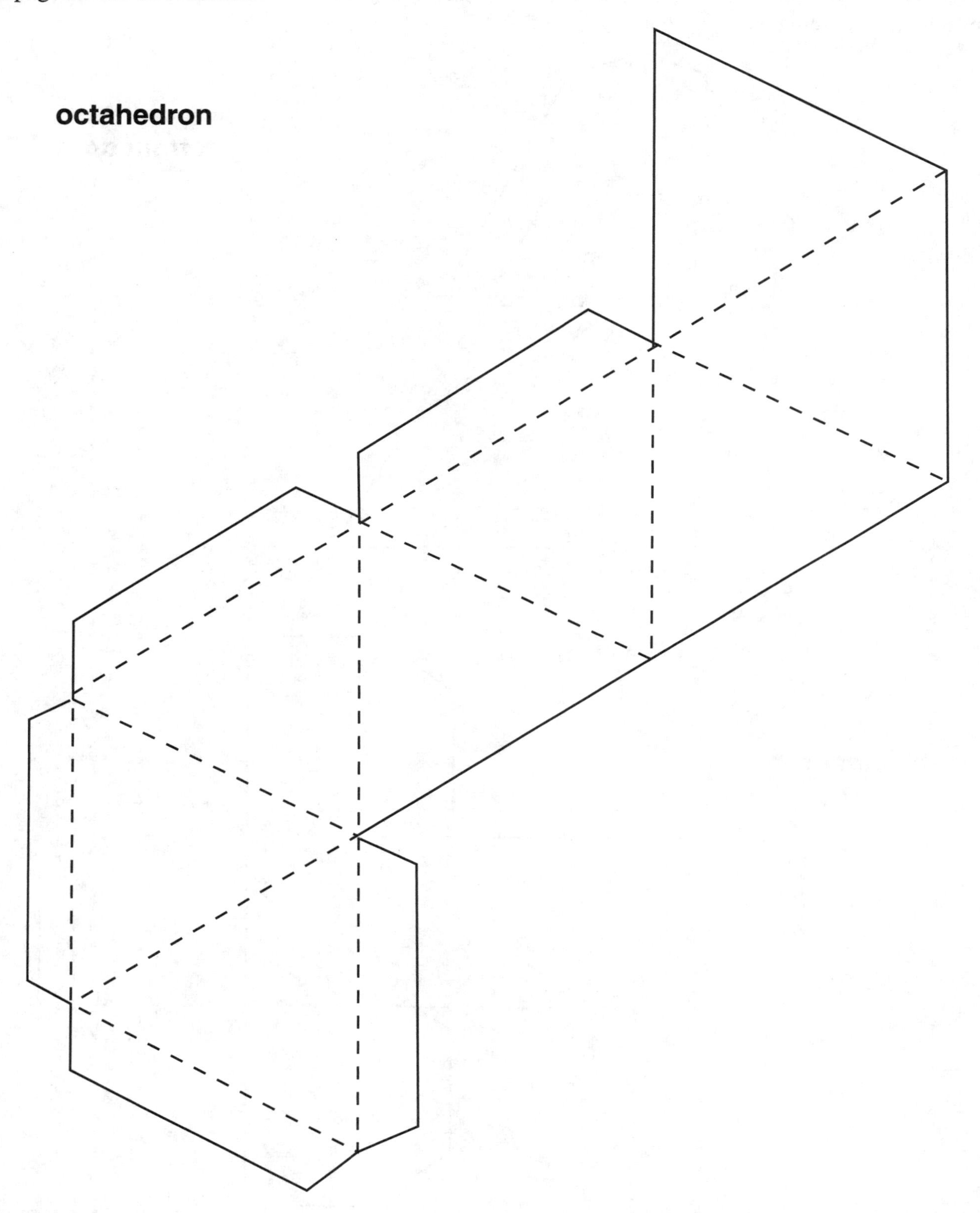

Polyhedron Patterns (cont.)

See page 75 for directions.

icosahedron

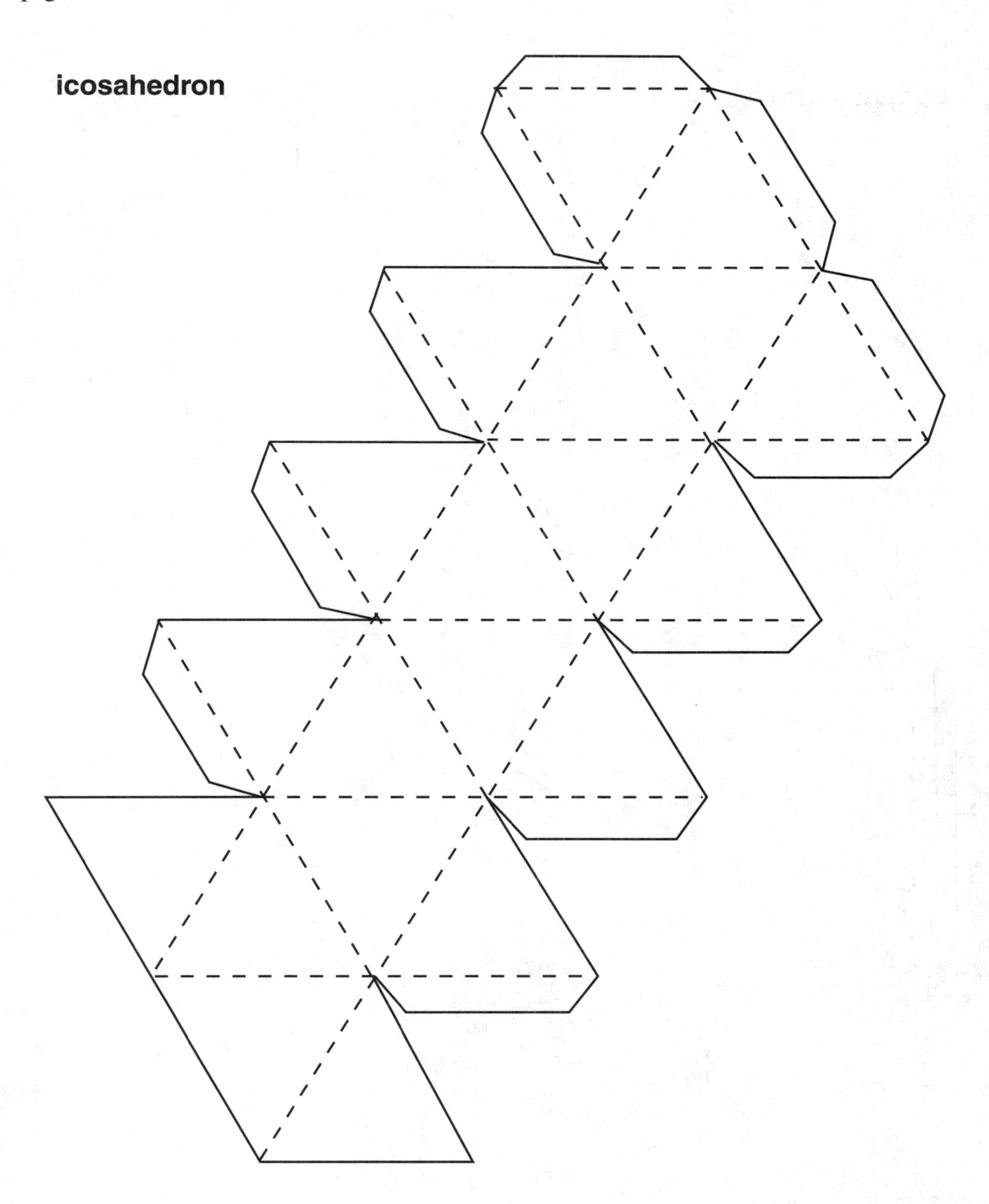

Polyhedron Patterns (cont.)

See page 75 for directions.

dodecahedron

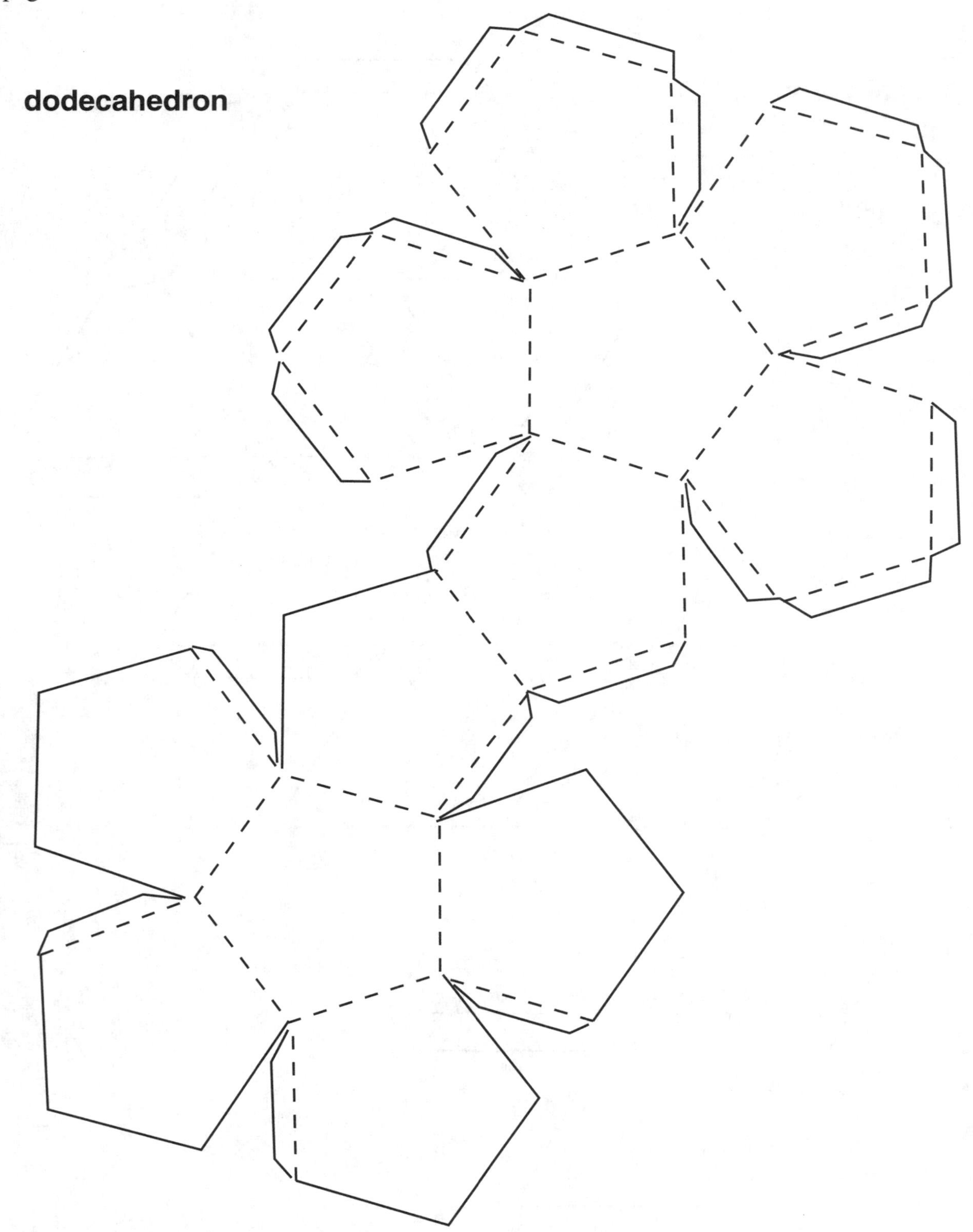

Finding Volume

The volume of a solid refers to the number of cubic units it contains. Although the formula for finding the volume of a solid varies depending on the type of solid, the same three dimensions (length, width, and height) are involved. Below are the formulas for finding the volume of a cube, a rectangular prism, and a cylinder. Can you see how each formula was derived? (Hint: Each volume is found by multiplying the area of the figure's base by the height.) Study the figures and the formulas and then use this information to solve the problems on page 80.

Type of Solid	Illustration	Formula	Notes
Cube	s	$V = s^3$	s = length of an side (or height of cube) Area of base = s^2 When you multiply this (s^2) by the height (s^2), the volume becomes s^3.
Rectangular Prism	h B	$V = B\,h$	B = area of base (or length times width) h = height of prism
Cylinder	h r	$V = \pi r^2 h$ or $B\,h$	B (circle) = area of circle (or πr^2) h = height of cylinder

Finding Volume (cont.)

Directions: Use the information on page 79 to solve the following problems. Use the space provided to show your computations.

1. The Petersons and the Andrews families both have swimming pools. What is the volume of each pool? Which pool holds the most water?

Petersons' Pool

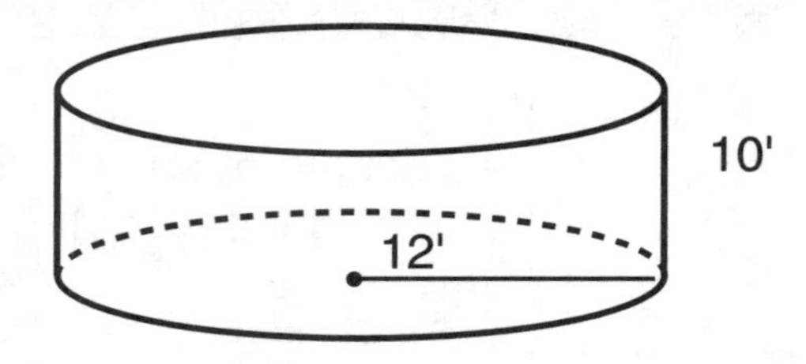

Andrews' Pool

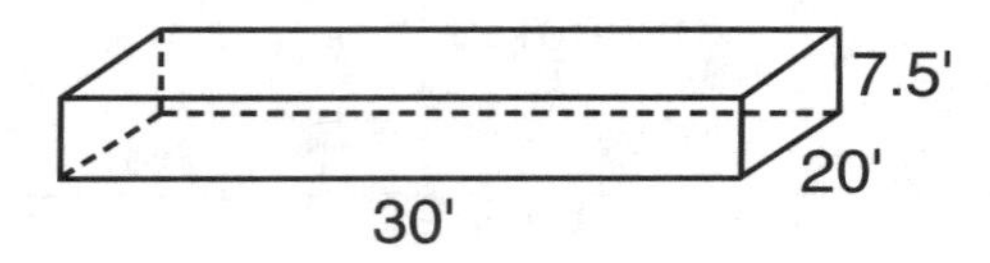

Computation

2. If you were to build the following block tower using solid blocks with the dimensions shown, how many cubic centimeters would it represent?

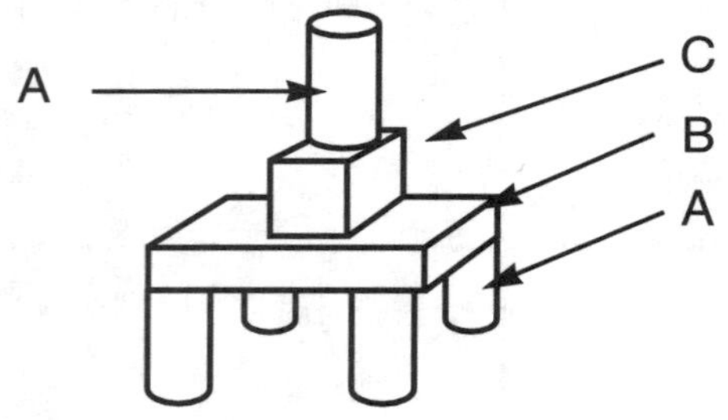

block A (cylinder)
r = 2 cm h = 10 cm
block B (rectangular prism)
base length = 25 cm
base width = 10 cm
h = 5 cm
block C (cube)
e = 6 cm

Computation

